This wise and lucid treatise explaining the promise and the hazards of alternative hedge fund strategies should be required reading for both investors and students of financial engineering.

*-**Burton G. Malkiel**, author of* A Random Walk Down Wall Street, *50th anniversary edition.*

The book covers both the classical works and recent advances in quantitative aspects of hedge fund investing. Broad in scope and commendable in erudition, the book easily earns a prime spot on hedge fund allocator's proverbial book shelf. Particular focus is rightfully paid to the portfolio construction aspects of hedge fund investing, alpha/beta separation of hedge fund returns, and persistence of those returns more generally. Personal stories covered among other material are a nice touch and make the whole book even more fun to read. In all, a welcome and long overdue addition to the professional literature on this thorny but relevant subject matter.

*-**Alexander Rudin**, Ph.D., Global Head of Multi-Asset and Fixed Income Research as State Street Global Advisors.*

Molyboga and Swedroe have produced a comprehensive guide to the theory and practice of hedge fund investing. From strategy and manager selection to portfolio construction, the book offers rigorous and pragmatic advice. I wish this book existed when I was starting out.

*-**Tobias Carlisle**, Managing Director, Acquirers Funds.*

This book covers a vast range, from classic topics of hedge fund performance sources, biases, and persistence, to smartly constructing hedge fund portfolios, to newer topics like diversity. Wonderful practitioner and expert interviews bring further flesh and color to the subject. I wish I had read this before I wrote my own books.

*-**Antti Ilmanen**, Principal, AQR Capital.*

Molyboga and Swedroe provide a comprehensive and insightful guide to quantitative investing in hedge funds. They explain carefully and lucidly all the technical details of this important area of investing. For each topic, they provide an excellent account of both the empirical evidence and the theory, based on results from the most recent academic research. This book provides the definitive cutting-edge guide for graduate students and investment professionals who wish to acquire a broader and deeper understanding of hedge fund investing. If you are going to read one book on hedge funds, you should read this one.

*-**Raman Uppal**, Professor of Finance, EDHEC Business School.*

The title doesn't do this book justice. It's about much more than quantitative hedge fund investing. It also discusses general manager and factor selection, whether performance persists, risk parity vs. traditional investing, even cutting edge topics like machine learning and important less quantitative topics like inclusion and diversity. In particular the interview section of the book was exceptionally informative save the one negative being that I was not an interviewee :) We will correct that in the next edition. The authors tackle this wide range of topics with their typical thoroughness and insight, and I recommend this book wholeheartily if you're interested in the titled subject or just good investing in general.

*-**Cliff Asness**, Managing and Founding Principal, AQR Capital Management.*

A very thorough guide on hedge funds from a refreshing allocator's perspective. Unlike the many books on hedge funds that take the fund manager's point of view and inevitably fall short on details for obvious secrecy reasons, this book gets deep into the data and modeling. The allocator's perspective, and this book in particular, would be my choice for a business school course on hedge funds.

-**Michael W. Brandt**, *Kalman J. Cohen Professor of Finance, Fuqua School of Business, Duke University.*

Finally, a comprehensive guidebook to systematic hedge fund investing that bridges the gap between academic research and the real-world practice of asset management. The authors do a great job de-mystifying popular trading strategies and the statistical tools behind them, zeroing in on the key problem: distinguishing investing skill from luck is fiendishly hard. The "human" side of the book in the form of interviews with practitioners exhibiting a wide diversity of backgrounds and experiences brings a refreshing new perspective on the opaque world of hedge fund investing.

-**Nikolai Roussanov**, *Moise Y. Safra Professor of Finance, Wharton Business School, University of Pennsylvania.*

I learned a lot from this book. There are cautionary histories of hedge funds' successes and failures, an impressively thorough review of the research on hedge fund performance, and engaging personal stories of hedge fund managers. Each chapter is followed by succinct "key takeaways." My grand takeaway is that investing in hedge funds is complex enough that I would not attempt to find a select set of them for my portfolio without the help of a wise, knowledgeable, and trusted advisor.

-**Edward Tower**, *Professor of Economics, Duke University.*

Hedge funds benefit from a mystique supported by perceptions of exclusivity and outsize performance available only to institutions and wealthy investors. This book takes a much-needed clear eyed approach to evaluating the portfolio value of hedge funds. Backed by dozens of academic studies, the authors provide a realistic evaluation of the headwinds faced by hedge funds hoping to provide value in a market filled with smart traders and barriers to persistence. I know of no other book that provides an equally exhaustive evaluation of the methodologies used to evaluate whether hedge funds are able to improve performance, and whether skilled advisors can select managers that provide value.

-**Michael Finke**, *Frank M. Engle Chair of Economic Security at The American College.*

As someone who teaches quantitative investing, this book is a wonderful resource for practical insight on quantitative methods in investing. Covering a wide range of topics that include portfolio construction, performance evaluation (and its biases), discretionary versus systematic funds, and newer topics on diversity, this book offers a wealth of information and tools for applying quant methods in finance. A wonderful resource for students and practitioners.

-**Toby Moskowitz**, *Dean Takahashi Professor of Finance at Yale University and AQR Principal.*

Well-researched and easy to read, this book is a must-read for all investors considering alternative investments. Grab a copy!

-**Wesley R. Gray**, *PhD, CEO of Alpha Architect.*

Your Essential Guide to Quantitative Hedge Fund Investing

Your Essential Guide to Quantitative Hedge Fund Investing provides a conceptual framework for understanding effective hedge fund investment strategies. The book offers a mathematically rigorous exploration of different topics, framed in an easy to digest set of examples and analogies, including stories from some legendary hedge fund investors. Readers will be guided from the historical to the cutting edge, while building a framework of understanding that encompasses it all.

Features

- Filled with novel examples and analogies from within and beyond the world of finance.
- Suitable for practitioners and graduate-level students with a passion for understanding the complexities that lie behind the raw mechanics of quantitative hedge fund investment.
- A unique insight from authors with experience of both the practical and academic spheres.

Your Essential Guide to Quantitative Hedge Fund Investing

Marat Molyboga
Larry E. Swedroe

CRC Press is an imprint of the
Taylor & Francis Group, an **informa** business
A CHAPMAN & HALL BOOK

Designed cover image: Shutterstock Images

First edition published 2023
by CRC Press
6000 Broken Sound Parkway NW, Suite 300, Boca Raton, FL 33487-2742

and by CRC Press
4 Park Square, Milton Park, Abingdon, Oxon, OX14 4RN

CRC Press is an imprint of Taylor & Francis Group, LLC

Library of Congress Cataloging-in-Publication Data

Names: Molyboga, Marat, author. | Swedroe, Larry E., author.
Title: Your essential guide to quantitative hedge fund investing / by Dr.
Marat Molyboga and Larry E. Swedroe.
Description: First edition. | Boca Raton, FL : Chapman & Hall | CRC Press,
2023. | Includes bibliographical references and index.
Identifiers: LCCN 2022061379 (print) | LCCN 2022061380 (ebook) | ISBN
9781032006963 (hardback) | ISBN 9780367776091 (paperback) | ISBN
9781003175209 (ebook)
Subjects: LCSH: Hedge funds. | Investment analysis. |
Investments--Mathematical models.
Classification: LCC HG4530 .M645 2023 (print) | LCC HG4530 (ebook) | DDC
332.64/524--dc23/eng/20221222
LC record available at https://lccn.loc.gov/2022061379
LC ebook record available at https://lccn.loc.gov/2022061380

ISBN: 978-1-032-00696-3 (hbk)
ISBN: 978-0-367-77609-1 (pbk)
ISBN: 978-1-003-17520-9 (ebk)

DOI: 10.1201/9781003175209

Typeset in Latin Modern font
by KnowledgeWorks Global Ltd.

Publisher's note: This book has been prepared from camera-ready copy provided by the authors.

Larry dedicates this book to his wife Mona. Whatever he has been able to accomplish has been enabled by her persistent and unquestioning support. Walking through life with her has truly been a gracious experience.

Marat dedicates this book to his kids, Olivia, Gabriel, and Maya, and his precious wife Julia, his better half and the love of his life, who motivates him to become the best version of himself.

Contents

Foreword

Having coauthored several research papers with Marat Molyboga, I was honored when he asked me to write a foreword for his book on quantitative hedge fund investing that he coauthored with Larry Swedroe. At its most basic level, quantitative hedge fund investing is simply about defining and then systematically following a set of rules that produce diversified portfolios that improve the efficient frontier for investors. However, once you define a strategy, questions immediately arise, such as: Did it make money in the past and why? Are there logical risk- or behavioral-based explanations for why we should believe it will continue to do so in the future? Are the excess returns only paper profits, or do they survive all implementation costs?

One classic example of a quantitative hedge fund strategy is time-series momentum, going long assets with positive returns in the recent past and shorting those with negative returns. There are many others, and in fact, that in and of itself can be a bit of a problem. How do you successfully identify strategies that will work in the future, and how do you identify managers who can implement these strategies efficiently?

Here's where Marat and Larry thankfully come in, providing a useful guide to understanding quantitative hedge fund investing and doing so in such a way that a non-super-geek interested reader can benefit. At the same time, in the appendices, they provide the quantitative details that the super-geeks need to implement a well-thought-out investment plan. These are very important, but not easy tasks that they do superbly. This is why their new book, *Your Essential Guide to Quantitative Hedge Fund Investing* is such a significant contribution.

As you read the book, you will find many important issues where Marat and Larry provide insightful commentary to demonstrate their appreciation for the complexities of hedge fund investing, including the manager selection and portfolio construction processes. Here are just some of the important matters discussed in the book that I believe readers will appreciate.

Hedge fund investing is generally not well understood. The book provides a concise, evidence-based rebuttal of common hedge fund myths. That section alone will help any hedge fund investor avoid some of the costliest mistakes (such as performance chasing or over-exposure to the stock market).

While equity or mutual fund researchers can comfortably rely on bias-free returns from the Center for Research in Security Prices (CRSP), hedge fund data must be carefully sourced and scrubbed. Marat and Larry provide a detailed guide for creating and using a bias-free dataset of hedge fund returns to improve performance.

Manager selection and portfolio construction are two critical topics for any investor. The book provides an excellent overview of cutting-edge portfolio management techniques and a complete framework for evaluating any quantitative approach with real-life constraints.

The last few chapters of the book are unusual for quantitative manuscripts. They share experiences and suggestions of accomplished hedge fund managers and investors who kept on searching for answers in their respective areas despite obstacles and setbacks. Their stories also have a common thread of diversity, inclusion, and fulfillment from helping others. I believe their stories will inspire young people to pursue their dreams, even if the odds may be stacked against them.

I encourage you to read this book and appreciate its broad scope and many lessons for developing a successful quantitative hedge fund investment strategy—it would make an excellent textbook for graduate students or serve as a guide for investment professionals. I am confident that you will enjoy your journey through quantitative hedge fund investing as much as I did.

Frank J. Fabozzi
Professor of Practice, Carey Business School, John Hopkins University
Editor, *Journal of Portfolio Management*

Preface

"It's good to learn from your mistakes. It's better to learn from other people's mistakes."—Warren Buffet.[a]

"Knowledge comes, but wisdom lingers. It may not be difficult to store up in the mind a vast quantity of facts within a comparatively short time, but the ability to form judgments requires the severe discipline of hard work and the tempering heat of experience and maturity."—Calvin Coolidge.[b]

As the director of research at Efficient Capital Management for more than two decades, I have enjoyed building customized multi-manager solutions for investors. It has not been an easy journey for me. When I joined Efficient Capital with an Applied Mathematics degree in 2001, I thought that quantitative portfolio management of hedge funds was easy because *modern portfolio theory* conveniently simplified the problem down to *Sharpe ratio* maximization. All I had to do was to estimate the vector of expected returns of the portfolio constituents and their covariance matrix. How hard could it be with years of daily returns for portfolio constituents and a huge library of statistical methods widely available?

However, as I was going through a list of statistical approaches and widely accepted portfolio management techniques, I kept seeing the same pattern—very few worked out-of-sample. And when my colleague Grant Jaffarian suggested risk-based approaches for portfolio construction, I was quick to dismiss his advice as one not based on financial theory. Unfortunately, it took me a long time to recognize that financial theory didn't provide easy answers because it often underestimated the role of luck and overestimated the information contained in historical performance.

[a]Warren Buffet is one of the most successful investors in the world.
[b]Calvin Coolidge was the 30th U.S. president.

I wanted to write this book for two reasons. First, I have been fortunate to work and collaborate with brilliant investment professionals and academics who passionately approach the topic of hedge fund investing with both scientific curiosity and rigor. They have shown me that quantitative hedge fund investing is full of widely accepted myths that must be debunked, and fascinating and challenging problems that can be successfully solved to benefit investors. Therefore, I wanted to write this book to honor those people and give back to others.

The second reason is a little selfish. I am very passionate about learning, and I was excited about this incredible opportunity to attain a deeper level of my own understanding of some of the most challenging and exciting topics of manager selection and portfolio construction. As the Roman philosopher Seneca said: "While we teach, we learn."

I approached Larry to help me write this book because he has dedicated his life to educating investors about evidence-based investing. In all of his books, Larry has been successful at distilling hundreds of heavy academic papers into pragmatic toolboxes of research-based investment ideas and decisions.

Together we wrote this book to help all investors better understand and manage hedge fund portfolios. Although many of the topics covered in our book are highly technical, we wanted to make it accessible to any hedge fund investor who is interested in evidence-based investing. To accomplish this objective:

- We were intentional about choosing simplicity over technical precision.
- We italicized the first use of technical financial, statistical, or machine learning terms and provided their definitions in a glossary.
- We summarized key takeaways at the end of each chapter.
- We included a list of all referenced academic papers at the end of each chapter.

We also wanted our book to be useful for the quants and graduate students who are interested in the highly technical aspects of investing that are required when implementing and testing sophisticated portfolio management approaches. Thus, we included four appendices that provide in-depth technical descriptions and implementation steps for most of the approaches covered in the book.

Larry and I want to take you on a journey into the fascinating world of hedge fund investing. Our book is not just about investing in quantitative hedge funds. It uses quantitative tools to help hedge fund investors invest in both systematic and discretionary hedge funds. We also hope that hedge fund managers will benefit from the techniques discussed in the book to improve their portfolios. We have the ambitious goal of guiding you through the process of building a hedge fund portfolio for investors, both individual and institutional, by following rigorous steps of determining which building blocks you can rely on and thoughtfully putting them together.

We start in Chapter 1 by providing an introduction to hedge funds and debunking several common myths: hedge fund investing and selection of top performers are easy; hedge funds hedge; active and socially responsible hedge funds outperform; and investors benefit from hedge funds identifying undervalued stocks. In Chapter 2, we pose essential research questions about hedge funds and discuss hedge fund databases and their inherent biases that create issues for empirical research.

In Chapter 3, we provide a thorough overview of the essential aspects of manager selection including quantitative and qualitative analysis. We demonstrate that a gap exists between standard academic methodologies and industry needs and then propose a robust and implementable framework for evaluating manager selection techniques. We also describe factors and factor selection techniques for performance evaluation of hedge funds.

In Chapter 4, we provide a comprehensive overview of critical topics of predictive manager selection that include separation of luck from skill and performance persistence. This chapter also includes a framework for combining quantitative and qualitative factors within a Bayesian framework. Manager selection is a key topic in hedge fund investing. In selecting and allocating to a manager, you are not just buying a track record. Instead, you are buying an investment process that you believe will produce an attractive risk-return profile for your portfolio going forward. It is critical to assess the sustainability of this investment process.

In Chapter 5, we introduce a practical customizable framework for the evaluation of portfolio construction approaches. We describe the evolution of portfolio construction approaches from mean-variance optimization and its extensions to strategies that diversify risk across managers (risk-parity) and time (*volatility*-targeting). In chapter 6 we describe advanced portfolio construction techniques that are relevant for both hedge fund investors and managers. The chapter covers several machine learning approaches, two recent cutting-edge methods, and several interesting complementary nuggets.

Together, the first six chapters present the empirical evidence from hundreds of research studies published in peer-reviewed academic financial and mathematical journals. In Chapters 7–9, we include personal stories and experiences of expert hedge fund managers and investors. I have personally benefited from their valuable insights that are based on decades of quantitative research and reflection. We believe that you will as well. Chapter 7 provides a deep dive into four hedge fund strategies: trend following; machine learning; emerging markets; and sustainable investing. Chapter 8 covers the challenging topics of quantitative and qualitative manager selection and quantitative portfolio construction. Chapter 9 takes you on a journey that shows why and how diversity and inclusion—both of which are important to Larry and me—lead to better investment decisions. It also describes a gender gap in financial academia. Everything discussed in Chapters 7–9 represents personal opinions and should not be interpreted as investment recommendations.

In a well-known tale, Ernest Shackleton, the famed explorer of Antarctica, posted an ad in the newspaper: "Men wanted for hazardous journey, small wages, bitter cold, long months of complete darkness, constant danger, safe return doubtful, honor and recognition in case of success." While practical evidence-based hedge fund investing is not hazardous physically, it requires a lot of hard work with small victories and painful setbacks. We hope you will enjoy the journey through this book.

Acknowledgments

Marat thanks Mike Marcey, Chad Martinson, Lorent Meksi and other colleagues at Efficient Capital for their support, collaboration and friendship. Marat is grateful to Alma Molyboga, Ernest, and Trula Jaffarian for providing opportunities and challenging him to grow personally and professionally.

Both Larry and Marat thank Mike Going, Georgiy Molyboga, Nikolai Semtchouk, Kevin Townsend, and Dan Zhang for their useful feedback on the draft. They appreciate the insightful comments of Antti Ilmanen. Larry and Marat are also grateful to Adam Duncan, Mila Getmansky Sherman, Meredith Jones, Katy Kaminski, Christophe L'Ahelec, Asha Mehta, Heidi Pickett, Christina Qi, Riti Samanta, Nina Tannetbaum, Kay Wu, and Jasmine Yu for sharing their insightful comments, practical suggestions, and personal stories.

About the Authors

Dr. Marat Molyboga is the Chief Risk Officer and Director of Research at Efficient Capital Management, where he helps shape strategic priorities as a member of the Leadership Team. He began his career at Efficient in 2001 as a Research Analyst, working on fund selection and portfolio construction algorithms. Later, he spent several years developing and managing intraday trading strategies for a firm subsidiary before assuming his current risk management and research roles. His expertise is in hedge fund performance evaluation and portfolio construction, specializing in practical applications of rigorous academic research.

Dr. Molyboga graduated with high honors from Moscow State University with a Master's in Financial Mathematics and honors from the University of Chicago's Booth School of Business with an MBA in Finance, Economics, and Strategic Management. He earned a Ph.D. in Finance from EDHEC Business School. Dr Molyboga is a Chartered Financial Analyst (CFA) and an Adjunct Professor at the Illinois Institute of Technology Stuart School of Business, where he teaches empirical finance techniques to Ph.D. students.

His research papers have been published in many academic and practitioner-oriented journals, such as the *Journal of Portfolio Management*, *Journal of Alternative Investments*, *Journal of Futures Markets*, and *Journal of Financial Data Science*. He also enjoys presenting his work at various academic and industry conferences.

Larry E. Swedroe is the head of Financial and Economic Research, Buckingham Wealth Partners. In his role as chief research officer and as a member of the firm's Investment Policy Committee, Larry regularly reviews the findings published in dozens of peer-reviewed financial journals, evaluates the outcomes, and uses the result to inform the firm's formal investment strategy recommendations. He has had his own articles published in the *Journal of Accountancy*, *Journal of Investing*,

AAII Journal, *Personal Financial Planning Monthly*, *the Journal of Portfolio Management*, and *Wealth Management*.

Since joining the firm in 1996, Larry Swedroe has spent his time, talent, and energy educating investors on the benefits of evidence-based investing with. Larry's dedication to helping others has made him a sought-after national speaker. He has made appearances on national television shows airing on NBC, CNBC, CNN, and Bloomberg Personal Finance. Larry is a prolific writer, contributing regularly to multiple outlets, including Advisor Perspectives, The Evidence Based Investor, and Alpha Architect.

Larry is a prolific writer, contributing regularly to multiple outlets, including *Advisor Perspectives*, *Wealth Management*, and *Alpha Architect*. Larry was also among the first authors to publish a book that explained the science of investing in Layman's terms, *The Only Guide to a Winning Investment Strategy You'll Ever Need*. The second edition was published in 2005. He has since (co)-authored 16 more books including, *Your Complete Guide to Factor-Based Investing* (2016), *Reducing the Risk of Black Swans* (2018), *The Incredible Shrinking Alpha* (2020), *Your Complete Guide to a Successful and Secure Retirement* (2021), and *Your Essential Guide to Sustainable Investing* (2022).

CHAPTER 1

Introduction to Hedge Funds

"Science must begin with myths, and with the criticism of myths."—Karl Popper.[a]

"It's not what you don't know that kills you, it's what you know for sure that ain't true."—Mark Twain.[b]

We start our journey in this chapter by providing an introduction to hedge funds and debunking several prevalent myths about hedge fund investing.

1.1 AN INTRODUCTION TO HEDGE FUNDS

The hedge fund industry is a byproduct of the article *Fashion in Forecasting* written for *Fortune* magazine by Alfred Winslow Jones in 1949. The article included this catchy subtitle: *"Stock market behavior as interpreted by the 'technicians' of statistics, charts, and trends. A report on the rising competitors of the Dow Theory, whose very popularity may have impaired its own usefulness."* It went on to describe Jones as a sociologist by profession.[1] As research for the article, Jones investigated a dozen methods used by market technicians such as James Hughes who relied on a two-day moving average of market advances minus declines to infer public participation in the market, Mansfield Mill who tried to

[a]Karl Popper is one of the most influential philosophers of science in the 20th century.

[b]Mark Twain is a famous American writer.

DOI: 10.1201/9781003175209-1

capture trends in the markets by calculating daily market-wide dollar gains and losses normalized by daily volume, and Nicholas Molodovsky who traded based on his formulaic confidence index. Jones was careful to contrast the systematic approaches that were based on detailed calculations of price-time-volume relations and the "wonder systems" exemplified by one that made market predictions based on the outcomes of the Harvard-Yale football games.

As Carol Loomis stated in his famous 1966 *Fortune* article *The Jones Nobody Keeps Up With,* it was his research for the *Fashions in Forecasting* that convinced Jones that he could make a living in the stock market.[2] He formed a general partnership that earned 17.3 percent during the first year employing his new "hedge" idea. Loomis described how Jones selected the "right" stocks to buy and sell short and how he chose the amount of risk to take. Jones' performance was spectacular. Over the period 1955–65, his hedge fund returned 670 percent, far outperforming all mutual funds, of which the leading performer was the Dreyfus Fund which was up 358 percent.

Today, Jones' long-short (also referred to as "equity hedge") approach is one of the most common types of hedge fund investing, representing about 5 percent of total hedge fund assets. It typically involves taking long positions in equities considered undervalued by a hedge fund manager, and taking short positions in equities considered overvalued. Some of the standard techniques used in long/short equity trading include:

- The value-based fundamental approach of Benjamin Graham;
- Quantitative (systematic) techniques based on well-known risk premia such as momentum, value, and quality;
- Sector-specific methods that attempt to leverage expertise in a niche sector such as pharmaceuticals or technology;
- Activism.

Among the hedge fund managers who represent this segment, David Einhorn, founder of Greenlight Capital, has been under scrutiny from regulators and the media for his trading ideas. In 2013, he was included in the *Time* magazine's *100 most influential people in the world.* In his August client letter, David Einhorn described his trading style: "Our investing style is not a closet index of long value and short growth. We look for security-specific differences of opinion and hope to capitalize on being right and the market eventually seeing it our way."

His book *Fooling Some of the People All of the Time: A Long Short (and Now Complete) Story* describes Einhorn's involvement in bringing to light the fraudulent practices of Allied Capital. On May 15, 2002, Einhorn gave a speech at a charity conference named after Ira Sohn, a *Wall Street* professional who died of cancer at the age of 29, benefiting Tomorrow's Children's Fund. The speakers, who over the years have included famous investors such as Paul Tudor Jones, Carl Icahn, and David Tepper, contribute to the event by sharing compelling investment ideas. Einhorn started by discussing his track record of success in shorting stocks that had returned approximately 30 percent per annum. Then he outlined issues faced by Allied Capital, a Registered Investment Company publicly traded on the New York Stock Exchange.[3] The impact of his speech was so dramatic that Allied Capital's stock immediately fell 20 percent. After the six-year investigation of both Allied Capital and Einhorn, the *Securities and Exchange Commission* (SEC) determined that Allied Capital broke securities laws related to valuation practices and record keeping.[4]

Another exciting example of hedge fund investing is the battle of titans, Bill Ackman on the short side and Carl Icahn and Dan Loeb on the long side, for Herbalife. The war started on December 20, 2012, with Bill Ackman, the founder and CEO of Pershing Square Capital Management, giving the presentation *Who Wants to Be a Millionaire,* in which he denounced Herbalife for being a pyramid scheme that needed to be investigated and shut down by the regulators. As a result, Herbalife fell from \$42.50 to \$26.

For Dan Loeb, the CEO of Third Point, this drop presented an opportunity to buy Herbalife shares at a bargain price. On January 9, Loeb filed a report with the SEC announcing that he had purchased approximately 8.24 percent of the company, making him the second-largest shareholder. About a week later, Carl Icahn, another prominent hedge fund manager, started acquiring shares of Herbalife and appeared on Bloomberg TV to make his position public. On February 14, Icahn announced his 12.98 percent stake in Herbalife and shared his plans to meet with its management to discuss changes that would benefit the shareholders. Interestingly, Dan Loeb announced in late January that he had previously made money shorting Herbalife, but he closed his positions earlier in the year.

The battle lasted for five years and ended with Bill Ackman exiting his position in 2018, having lost almost $1 billion. In contrast, Carl Icahn accumulated a 26 percent stake in the company and made almost $1 billion in profits.[5] After Bill Ackman closed his position, the SEC fined Herbalife more than $122 million for violations of the Foreign Corrupt Practices Act and for making corrupt payments to Chinese government officials.[6] The battle was described by Scott Wapner in his book *When the Wolves Bite: Two billionaires, One company, and an Epic Wall Street Battle*, in which he shared fascinating details, such as Herbalife's CEO Michael Johnson ordering and receiving a top-secret 30-page report on Bill Ackman that read like a spy novel and included an in-depth psychological profile prepared by Dr. Park Dietz, a leading forensic psychiatrist.

If you are interested in learning about prominent funds from different segments of the hedge fund industry, we recommend that you consider researching the following list of strategies and funds:

- Alternative Risk Premia (Cliff Asness' AQR);
- Convertible Arbitrage (Kenneth Griffin's Citadel);
- Commodity Trading Advisors (Andrew Lo's Alphasimplex);
- Distressed Securities (David Tepper's Appaloosa Management);
- Equity Market Neutral (James Simons' Renaissance Technologies);
- *Emerging Markets* (Bill Browder's Hermitage);
- Fixed Income Arbitrage (Robert Merton and Myron Scholes' Long-Term Capital Management);
- Global Macro (George Soros' Soros Fund Management);
- Merger Arbitrage (John Paulson's Paulson Partners);
- Risk Parity (Ray Dalio's Bridgewater);
- Systematic Trading (Leda Braga's Systematica).

1.2 MYTHS ABOUT HEDGE FUNDS

Investors need to be able to separate facts from fiction about hedge fund investing. In this section, we debunk several myths: hedge fund investing and selection of top performers are easy; hedge funds hedge; active and socially responsible hedge funds outperform; and investors benefit from hedge funds identifying undervalued stocks.

1.2.1 Hedge Fund Investing Is Easy

Hedge funds are run by incredibly intelligent people with Ph.Ds. in advanced fields such as rocket science and string theory. The industry attracts some of the brightest minds, who are intrigued by the promise of massive monetary rewards in exchange for cracking the hidden code of the financial markets. The fundamental question is whether intellectual power is just a necessary condition for successful hedge fund investing or is it a sufficient condition. The question arises because there are two real challenges to success in the hedge fund industry: market efficiency and the importance of the relative, rather than absolute, skill level.

What so many people fail to comprehend is that in many forms of competition, such as chess, poker, or investing, it is the *relative* level of skill that plays the more important role in determining outcomes, not the *absolute* level. What is referred to as the "paradox of skill" means that even as the skill level rises, luck can become more important in determining outcomes if the level of competition is also rising.

In the July/August 2014 issue of the Financial Analysts Journal, Charles Ellis noted: "over the past 50 years, increasing numbers of highly talented young investment professionals have entered the competition.... They have more-advanced training than their predecessors, better analytical tools, and faster access to more information."[7] Legendary hedge funds, such as Renaissance Technologies, SAC Capital Advisors, and D.E. Shaw, hire Ph.D. scientists, mathematicians, and computer scientists. MBAs from top schools, such as Chicago, Wharton and MIT, flock to investment management armed with powerful computers and massive databases. The unsurprising result of this increase in skill is that the increasing efficiency of modern stock markets makes it harder to match them and much harder to beat them, particularly after covering costs and fees.

Market efficiency has proven to be an increasingly difficult obstacle to overcome. For example, the authors of the 2014 study *Conviction in Equity Investing* found that the percentage of mutual fund managers that demonstrated sufficient skill to overcome their costs had fallen from about 20 percent in 1993 to less than 2 percent by 2011.[8] Fama and French found a similar result in their 2010 study, *Luck versus Skill in the Cross-Section of Mutual Fund Returns.*[9] With that said, given the potentially huge rewards for discovering *arbitrage* opportunities, and the amount of brain and computer power available, it would be naive to think that there would never be arbitrage opportunities that could be exploited, at least temporarily. In fact, there are limits to arbitrage (due to the costs and risks of shorting) that allow anomalies to persist. One example of how hedge funds could theoretically prosper comes from the world of convertible bond arbitrage. A hedge fund operating in the *asset class* of convertible bonds might be able to buy a convertible bond, short the issuer's equity, and lock in a profit. Or, the fund manager might simultaneously go long the equity and short the convertible bond. In either case, a profit could be locked in without accepting any net exposure to the risk of the stock. Searching for these anomalies seems like a desirable proposition.

Unfortunately for hedge funds and their investors, the arbitrage process rapidly brings prices back into equilibrium. Purchasing the undervalued security raises its price, and shorting the overvalued one lowers its price. This is the power of the efficient markets hypothesis, as expressed by economics professors Dwight Lee and James Verbrugge of the University of Georgia in the 1996 paper *The Efficient Market Theory Thrives on Criticism*:

"The efficient markets theory is practically alone among theories in that it becomes more powerful when people discover serious inconsistencies between it and the real world. If a clear efficient market anomaly is discovered, the behavior (or lack of behavior) that gives rise to it will tend to be eliminated by competition among investors for higher returns.... (For example) If stock prices are found to follow predictable seasonal patterns... this knowledge will elicit responses that have the effect of eliminating the very patterns that they were designed to exploitâĂęThe implication is striking. The more empirical flaws that are discovered in the efficient markets theory, the more robust the theory becomes. (In effect) Those who do the most to ensure that the efficient market theory remains fundamental to our understanding of financial economics are not its intellectual defenders, but those mounting the most serious empirical assault against it."[10]

The story of the most famous (and infamous) hedge fund, Long Term Capital Management (LTCM), is an example of just how powerful a force is market efficiency. The firm, founded by some of the brightest stars on Wall Street, attracted some of the top minds in academia, including two Nobel Laureates. The firm's strategy was to exploit market anomalies (mispricings). Unfortunately, the tyranny of the efficient markets, events, and their hubris conspired to overwhelm the assembled brainpower. As a result, investors lost billions of dollars. The failure of the fund even threatened the global financial system. Eventually, under Alan Greenspan, the Federal Reserve negotiated a lender bailout that allowed for an orderly unwinding of the fund.

Market efficiency is an enormous obstacle. As Rex Sinquefield, co-chairman of Dimensional Fund Advisors, pointed out, "Just because there are some investors smarter than others, that advantage will not show up. The market is too vast and too informationally efficient."[11] The result is that by the time you can identify a hedge fund that has successfully exploited an *anomaly*, the anomaly may have already disappeared.

The importance of the relative rather than absolute skill level further compounds the challenge. We can illustrate it using the example of chess. Chess is different from many other games, such as baseball or basketball, because of the lower impact of randomness. Although players can still make mistakes in chess, they don't have to worry about external factors such as wind speed or ball bounce.

The World Chess Championship in 2021 between Magnus Carlsen, the World Champion, and Ian Nepomniachtchi, the Challenger, illustrates the importance of relative advantage. Although Magnus won the match with a decisive score of 7.5-3.5, the turning point was the sixth game, a historic 136-move game, the longest game in world championship history, preceded by five draws. Each player spent almost six months preparing for the event, supported by a world-class team of "seconds" who spent thousands of hours developing new ideas. Magnus' team included Peter Heine Nielsen, Laurent Fressinet, Jan Gustafsson, Jorden van Foreest, and Daniil Dubov. Ian's team had Vladimir Potkin, Sergey Yanovsky, Peter Leko, and Sergey Karjakin. Each player also had access to state-of-the-art supercomputers. Ian heavily relied on the Zhores supercomputer from the Skolkovo Institute of Science and Technology in Moscow, which was capable of evaluating tens of millions of positions per second. Ian Nepomniachtchi highlighted the importance of the relative rather than absolute advantage: "You're more sure that your analysis is good when you see 500 million node positions than, say 100 million. In general, all the top players have access to something similar. And it's the chess engines, such as Stockfish and Leela Chess Zero, which are the main tools in helping us prepare. Everyone has those."[12]

It's also important to understand that while chess is a zero-sum game (for each winner there is a loser), investing is a less-than-zero-sum game because of the trading costs, *management and incentive fees*, operating expenses, and taxes. In that sense, investing is more like playing poker at the tables in a Las Vegas casino where the dealer is taking the house's share out of each pot. Therefore, the question becomes: Do hedge fund managers have enough relative skill to compensate for all the incremental costs and risks? The evidence is mixed. For example, Gaurav Amin and Harry Kat, authors of the 2003 study *Hedge Fund Performance 1990–2000: Do the "Money Machines" Really Add Value?* found that hedge funds have failed to offer superior risk-adjusted performance as stand-alone investments, although they can have a positive marginal contribution to portfolios of stocks proxied by *S&P 500* and, therefore, recommended allocating 10–20 percent to hedge funds.[13] Others, such as Robert Kosowski, Narayan Naik, and Melvyn Teo, in their 2007 study *Do Hedge Funds Deliver Alpha? A Bayesian and Bootstrap Analysis* showed that hedge funds produced positive net-of-fee alpha, on average.[14]

The 2021 study *The Hedge Fund Industry is Bigger (and has Performed Better) Than You Think* by Daniel Barth, Juha Joenvaara, Mikko Kauppila, and Russ Wermers contributed to the debate by supplementing hedge fund returns from commercial databases (publicly reporting funds) with the regulatory data for U.S. funds (non-publicly reporting funds) that do not report to any public database.[15] The authors discovered that non-publicly reporting funds delivered superior risk-adjusted performance relative to the publicly reporting funds. They also found strong empirical evidence of performance persistence among non-publicly reporting funds and little or no evidence of performance persistence among publicly reporting funds.

Another challenge that hedge fund investors have to overcome is that the publication of research leads to the transition from *alpha* (a source of excess return) to *beta* (exposure to a common trait or characteristic). For example, prior to the development of the *Fama-French three-factor* (*market beta*, size, and value) model in 1992, actively managed funds could produce higher returns than a *benchmark*, such as the *Russell 2000 Index* or the S&P 500 Index, by "tilting" their portfolio to either *small stocks* or *value stocks*, thus giving them more exposure to the size and value *factors* than the benchmark index. The fund would then claim that its outperformance was, in fact, alpha. Today, regression analysis would show that their outperformance was simply the result of a greater exposure to certain factors. In effect, what once was alpha had now become beta, or what is referred to as alternative risk premia that can be accessed with low cost index and other "passive" strategies that are systematic, transparent, and replicable. Other examples of factors that used to be sources of alpha, but are now considered simply as exposures to different betas, are size, value, momentum, profitability, quality, low volatility, term, default, and carry.

1.2.2 It Is Easy to Select Top-performing Hedge Funds

The influential 2004 paper *Mutual Fund Flows and Performance in Rational Markets* by Jonathan Berk and Richard Green presented a rational model of *active management* where fund managers with positive gross-of-fee alpha attract investor capital until the net-of-fee alpha disappears because of decreasing returns to scale.[16] This study highlights the challenge of finding active managers with superior out-of-sample performance in any investment field, including hedge funds. The 2022 study *Hedge Fund Flows and Performance Streaks: How Investors Weight Information* by Guillermo Baquero and Marno Verbeek demonstrated that flows into hedge funds are highly sensitive to performance streaks—investors chase recent performance—yet the funds chosen by investors fail to perform significantly better than the funds from which the investors divested.[17] The 2018 study *Alpha or Beta in the Eye of the Beholder: What Drives Hedge Fund Flows?* by Vikas Agarwal, Clifton Green, and Honglin Ren found that investor flows followed performance relative to the *Capital Asset Pricing Model.* However, that behavior didn't lead to better performance.[18] These empirical findings are consistent with the prediction of Berk and Green. The prediction of Berk and Green is a direct result of their key assumption of perfect competition for skilled hedge fund managers by the many capital providers. Since hedge fund managers also compete for capital and past performance is noisy, it is hard for managers to attract capital. Thus, only a portion of gains, rather than all gains from skill, should accrue to managers.

Another critical effect in hedge funds is the impact of flows into hedge fund categories or strategies as documented in the 2012 paper *The Life Cycle of Hedge Funds: Fund Flows, Size, Competition, and Performance* by Mila Getmansky and the 2009 paper *Crowded Chickens Farm Fewer Eggs: Capacity Constraints in the Hedge Fund Industry Revisited* by Oliver Weidenmueller and Marno Verbeek.[19,20] The latter study examined the evidence on more than 2,000 hedge funds over the period 1994–2006 to see if cash inflows and capacity constraints could explain the lack of persistence in performance. For hedge funds, cash inflows are not only a problem at the fund level but also at the strategy level. If there are anomalies that allow for the creation of alpha, cash flows will follow. As we have discussed, the very act of exploiting an anomaly, combined with the increased cash flows that follow and the competition from imitators, will cause the anomaly to shrink and perhaps eventually disappear. This phenomenon is known as "the tyranny of an efficient market."

Following are the conclusions of the 2009 study by Oliver Weidenmueller and Marno Verbeek:

- Inflows hurt small rather than large funds because the negative effect of being past an optimal size predominates.
- The increased competition—funds chasing similar investment opportunities—leads to a reduction of the average level of alpha. This effect applies to skilled and unskilled managers.
- Their findings confirmed the results of other studies that found both that inflows lead to worse future performance and that there is little support that performance is persistent in hedge funds.

Hedge funds not only suffer from the negative impact of cash inflows, they also face the problem that any alpha generating strategy suffers in performance when competitors follow similar strategies, and strategies become "crowded." The more capital is allocated to funds that follow similar alpha seeking strategies, the lower is the average alpha of each fund.

The aforementioned 2012 paper *The Life Cycle of Hedge Funds: Fund Flows, Size, Competition, and Performance* by Mila Getmansky is one of the most comprehensive studies of hedge fund performance. Getmansky used the bias-free dataset of 3,501 hedge funds from the Lipper TASS database to investigate the performance-flow relationships for individual funds and hedge fund categories, hedge fund competition within categories. She also investigated the optimal asset size problem for hedge funds from different hedge fund categories. Getmansky presented evidence of performance chasing by hedge fund investors and negative return to scale in hedge fund performance. However, similar to the 2009 study by Oliver Weidenmueller and Marno Verbeek, Getmansky also found evidence of competition among hedge funds within hedge fund categories that followed a similar pattern:

1. A strong performance of a hedge fund category is followed by investors' asset flows into the category.

2. A higher amount of assets increases the competition for the same limited opportunity set or alpha.

3. The reduction in alpha hurts the performance of all funds in the category and forces the hedge funds with marginal performance to liquidate.

The very act of exploiting market mispricings makes them disappear. It is the tyranny of market efficiency at work. Thus, the amount of alpha available to the industry isn't constant. We should logically expect that it should shrink over time. The aforementioned 1998 failure of LTCM, the largest hedge fund in the world at the time, provides a perfect illustration of this simple fact.

LTCM's strategy to arbitrage what it considered market mispricing produced spectacular returns in the early years, bringing in more and more assets to manage. However, tens of billions of dollars from competing firms began to chase the same spread opportunities LTCM had been pursuing. Thus, the size of the spreads it had been exploiting began to narrow, and profit opportunities diminished. To continue to earn the same returns for its investors, LTCM had to take on ever-larger positions and use more and more *leverage* to earn the same returns. At the beginning of 1998, the firm had equity of $4.7 billion and had borrowed over $124 billion to acquire assets of around $129 billion. It also had off-balance sheet derivative positions amounting to $1.25 trillion.

Leverage is a double-edged sword, magnifying both gains and losses. And the danger of using leverage is that you may have to be right all the time to be successful. The reason is that short-term losses may force investors to meet margin calls as the value of their collateral, on which the margin loan is based, shrinks. If you cannot meet the margin call, your collateral is liquidated in order to close the potentially profitable position. This lesson was one that LTCM either forgot or ignored (making the mistake of treating the highly unlikely as impossible).

Eventually, the markets went against LTCM. Previously, when its positions were smaller, the firm could hold on to the trades by coming up with additional collateral to meet the margin call. In this case, the size of the market's move and the amount of leverage deployed made meeting margin calls impossible. The firm had to liquidate positions at the worst possible time, further driving prices against itself as it unwound these positions. Eventually, the losses overwhelmed its ability to raise collateral and the banks called in their loans. The following insightful quote has often been attributed to John Maynard Keynes, perhaps the most famous economist of modern times: "The market can stay irrational longer than you can stay solvent."

The word of caution for investors is that whenever an investment strategy that is exploiting some market mispricing has become popular, it might be already too late to join the party. Even worse, as Bill Bernstein pointed out in his book *Skating Where the Puck Was,* when a strategy becomes popular, not only will it have low expected returns due to the crowding, but the investors are now "weak hands" which tend to panic at the first sign of trouble.[21] That leads to the worst returns occurring at the worst times when the correlations of all risky assets move toward one.

Several empirical studies show a decline in the average net-of-fee alpha of hedge funds. As shown in the 2021 study *Hedge Fund Performance: End of an Era?,* hedge fund performance declined over the 1997–2016 period.[22] For example, the percentage of funds with significantly positive Fung and Hsieh seven-factor alpha drops from 20 percent to 10 percent, whereas the percentage of funds with significantly negative alpha increased from 5 percent to roughly 20 percent. The authors investigated several potential explanations for the decline in performance and concluded that it was caused by increased regulation and central bank stimulus activity.

The 2022 paper *Anticipatory Trading Against Distressed Mega Hedge Funds* by Vikas Agarwal, George Aragon, Vikram Nanda, and Kelsey Wei showed the additional challenges that mega hedge funds face.[23] The authors discussed another explanatory factor for the underperformance of large hedge funds, one caused by a risk not understood by most investors. They began by noting that "the hedge fund industry provides an ideal setting for the best and brightest investment managers to leverage their investment ideas and be rewarded for investment success. The largest and most successful hedge fund managers are among the world's wealthiest people and achieve celebrity status. Therefore, perhaps not surprisingly, the trading strategies of such mega hedge fund (MHF) managers are heavily scrutinized by market participants. Public disclosures of MHFs' (hedge funds with more than $1 billion in assets under management) stock positions (mandated by regulation) are regularly discussed by the financial media and closely followed by competitors and copycat investors—their quarterly 13F filings being downloaded more than twice as often as those of non-MHFs. However, when MHFs suffer a setback or a surprising loss that forces them to liquidate assets, their need to liquidate is often known to other traders. This phenomenon has important implications for financial markets because predictable trading by distressed traders, especially large traders like MHFs, can be exploited by strategic traders in ways that further reduce liquidation values and impair price efficiency. Specifically, given the prospect of distressed selling by MHFs, other traders may rush to sell stocks in anticipation of negative return shocks resulting from MHFs liquidating a large position in response to margin calls or investor redemptions. These anticipatory trading activities can be intensified by the belief that trades by copycat investors that typically follow MHFs' investments would exacerbate any price impacts of liquidation by distressed MHFs."

Based on the above, the authors hypothesized that "front-running" trading behavior can lead to prices falling further below fundamental values, amplifying the distress of MHFs and causing even more significant losses. They sought to answer the following questions: "Do institutional investors trade in the same direction prior to the anticipated stock trades of distressed MHFs and, in this sense, engage in front-running? Does such anticipatory trading adversely impact distressed MHFs, as reflected in worse portfolio performance? Finally, are stocks that are held by distressed MHFs and targeted for front-running associated with greater price drops and reversals (for example, are such stocks more prone to prices deviating from their fundamental value?)." Their data sample covered the quarterly stock holdings of MHFs and other institutional investors over the period 1994–2018. They focused on distressed MHFs (their returns were both negative and ranked in the lowest quartile during the quarter) noting that relatively poor performance and losses can trigger redemptions from fund investors and/or margin calls on levered positions that force the MHF to liquidate large positions for loss. In addition, the authors noted that "due to their sheer large size, MHFs' trading activities can be expected to impact stock prices, motivating other institutions to trade ahead of distressed MHFs." Further, "MHFs' portfolio holdings are closely watched by other investors as evidenced by their quarterly 13F filings being downloaded more than twice as often as those of non-MHFs. Consequently, the market impact related to both anticipatory and copycat trading is potentially greater for stocks held by distressed MHFs as compared to distressed non-MHFs."

Following is a summary of their findings:

- MHFs account for about 25 percent of industry assets, use significant leverage, and more than half have significant lockup provisions.
- Distressed MHFs experience a much bigger blow in money flows following their poor performance relative to both non-distressed MHFs and distressed non-MHFs.
- There is significant predictability in selling by MHFs—existing holdings and past returns (momentum) predict trading.
- Institutional investors trade in the same direction as the anticipated trades of distressed MHFs. In anticipation of a 1 percent drop in stock ownership by all distressed MHFs next quarter, non-distressed MHFs reduced their stock ownership by 1.8 percent in the current quarter. The evidence of anticipatory trading is concentrated among institutions that arguably have greater discretion and incentive to engage in front-running, such as non-distressed hedge funds and mutual funds; other institutional types (e.g., banks, insurance companies, pension funds) showed no such front-running behavior.
- The evidence is strongest among front-running institutions with more resources and more patient capital (e.g., large funds, mutual funds with smaller flow volatility, and hedge funds with lockup provisions), and stocks most vulnerable to fire sales (e.g., illiquid stocks).
- Stocks that were expected to be more heavily sold by distressed MHFs exhibited greater abnormal short interest.
- The intensity of front-running predicts worse performance for MHFs during periods of distress. The economic magnitude was significant: a one *standard deviation* increase in front-running beta predicted 1.6 percent lower risk-adjusted (for the factors of market beta, value, and momentum) abnormal returns for long equity portfolios held by distressed MHFs over the following year relative to other MHFs—evidence consistent with distressed MHFs realizing lower liquidation values on their stock trades due to the anticipatory selling by other institutions.

- Stocks that were anticipated to be sold by distressed MHFs in the next quarter were associated with 1.7 percent lower abnormal returns during the current quarter. These return effects were only temporary because the same stocks earned positive abnormal returns over the following year (1.4 percent). The fact that the negative return effect subsequently reversed over future periods helps rule out the possibility that the negative abnormal returns reflected a deterioration in stock fundamentals or front-runners' stock picking skill; instead, the price effects most likely reflect temporary price pressure from anticipatory selling—return reversals were only significant among stocks that were heavily sold by other institutions during the current quarter.

The authors concluded that their findings have important implications for market efficiency, not only because such front-running can temporarily destabilize market prices, but also because it can adversely impact MHFs that may have the greatest capacity for informed trading. Their findings also reveal another mechanism that can contribute to diseconomies of scale in active management. In addition, they provide yet another explanation for why it is challenging to select top-performing hedge fund managers.

1.2.3 Hedge Funds Hedge

As investors build their portfolios, they need to understand how different holdings in their portfolios are exposed to *systematic risk* factors, such as the stock market (market beta). Since the term "hedge fund" seems to imply hedging, some investors may conclude that the performance of hedge funds is not correlated to the stock market. This myth is debunked in the study by Clifford Asness, Robert Krail, and John Liew *Do Hedge Funds Hedge?*, the *Journal of Portfolio Management* article that won the annual Bernstein Fabozzi/Jacobs Levy Best Article Award in 2001.[24] The authors began by examining the econometric properties of the monthly returns of hedge funds for the period 1994–2000. They reported a positive *serial correlation* that was likely driven by their holdings of illiquid exchange-traded securities and difficult-to-price over-the-counter securities, which is consistent with the findings of the 2004 paper *An Econometric Model of Serial Correlation and Illiquidity in Hedge Fund Returns* by Mila Getmansky, Andrew Lo, and Igor Makarov.[25] The authors showed that the nonsynchronous data approach of Elroy Dimson, Myron Scholes, and Joseph Williams for estimating betas as summed betas of regressions of returns on the contemporaneous and lagged market returns substantially increases the estimates.[26,27] For example, the beta of the aggregate hedge fund index to the S&P 500 index more than doubles from 0.40 to 0.84. This finding suggests that the performance of hedge funds is heavily influenced by the performance of the stock market.

The authors also investigated whether the adjusted betas are asymmetric by comparing betas estimated for months with positive S&P 500 returns (up markets) to those estimated for months with negative S&P 500 returns (down markets). For most hedge fund strategies, adjusted betas for up markets are higher than those for down markets. For example, the up market beta for fixed income arbitrage is equal to 0.08, whereas its down market beta is equal to 0.7. This finding is striking because it is opposite to what investors would expect from a hedged investment. The only exception is managed futures that has a small positive beta of 0.09 during up markets and a negative beta of −0.40 during down markets—indicating that this hedge fund strategy tends to provide downside protection during periods of market distress.

The authors of the 2016 study *Hedge Fund Tail Risk: an Investigation in Stressed Markets* contributed to the literature on hedge funds by examining the risk and performance of a portfolio of hedge funds.[28] Using three measures of risk (volatility, value-at-risk, and expected shortfall), they constructed a model allowing them to accurately predict portfolio volatility during normal times and capture in a realistic way stress moves during crisis periods. Their model is consistent with the empirical observation that returns in many financial markets are characterized by distributions with fat left tails.

Their study, which covered the period 1994–2011, used data for eight equal-weighted equity-related strategy indices from the Dow Jones Credit Suisse Hedge Fund database. The data is net of all fees and accounts for survivorship bias. Following is a summary of their key findings:

- Hedge funds contribute to the left-tail risk of a portfolio, which appears during crises. Most hedge fund strategy indices exhibit significant negative *skewness* and *excess kurtosis*.
- The contributions to tail risk are not limited solely to market beta. Hedge fund strategies are also exposed to other common

risk factors well-documented in the literature, such as size, value, momentum, credit, term, volatility, and the dollar.

- Factors that contribute to tail risk include *liquidity* risk and credit risk.
- Emerging markets exposure makes the greatest contribution to tail risk.
- During crises, even strategies such as market neutral and convertible bond arbitrage contribute to tail risk. Although over the full period, they slightly reduce the tail risk.

The authors concluded: "The natural ability of some hedge fund strategies to be hedgers to the total portfolio risk disappears during crisis periods." They wrote: "This is important especially during crisis periods, as investors seek *diversification* and hedging benefits from hedge funds."

1.2.4 Active Hedge Funds Outperform

Do more active hedge fund strategies produce better performance than the less active ones? The 2016 study *Returns to Active Management: The Case of Hedge Funds* used the Carhart four-factor (market beta, size, value, and momentum) model as the basis for comparison to investigate whether more active hedge funds provided higher risk-adjusted returns.[29] The authors used a novel but an intuitive approach to proxy hedge fund activeness. They first estimated the dynamics of factor loadings on a standard benchmark model and then used time-varying estimates of risk exposures to construct a measure of activeness for each fund. Their database included a large sample of 2,323 live and dead U.S. equity long/short hedge funds covering the period 1994–2013.

The authors hypothesized: "A priori, it is not clear whether the after-fee performance of the more active funds should exceed those of the less active funds. Fund managers that have skills in the selection of securities may follow a buy-and-hold approach, while those who have skills in timing various segments of the market may follow a more active strategy. However, if both active and less active fund managers are equally skilled, or if markets are efficient, then, because of the transaction costs, we should expect to see lower performance on the part of the active managers."

Following is a summary of their findings:

- Hedge funds tend to have positive exposures to the size factor, negative exposure to the value factor, and positive exposure to the momentum factor.

- When raw returns measure performance, a *monotonic*, positive relationship exists between activeness and performance—the more active the fund, the higher the raw return. However, the highly active funds' returns are more volatile than the least active funds' returns.

- The relationship between activeness and risk-adjusted return is negative for low to moderate levels of activeness. As activeness increases, the relationship between activeness and mean alpha turns flat with some notable fluctuations. Finally, for relatively high levels of activeness, there is a noticeable positive relationship between activeness and mean alpha. This relationship turns positive only at the highest levels of activeness.

The authors concluded: "If any, only a handful of active managers are successful in generating positive risk-adjusted returns for their funds." They added: "A more active hedge fund investment strategy is not associated with higher risk-adjusted returns."

1.2.5 Socially Responsible Hedge Funds Outperform

As institutional investors such as pension funds, sovereign wealth funds, and university endowments embrace socially responsible investing (SRI), investment managers can signal their commitment to responsible investment by signing the United Nations Principles for Responsible Investment (PRI). Attesting to the spectacular growth in investor interest in responsible investment, assets under management of PRI signatories had grown from $6.5 trillion in 2006 to $86.3 trillion in 2019.

PRI signatories are expected to adhere to the following six principles:

- To incorporate environmental, social, and governance (ESG) issues into investment analysis and decision-making processes.

- To be active owners and incorporate ESG issues into ownership policies and practices.

- To seek appropriate disclosure on ESG issues by the entities in which they invest.

- To promote acceptance and implementation of the principles within the investment industry.
- To work together to enhance effectiveness in implementing the principles.
- To report their activities and progress towards implementing the principles.

Does a commitment to SRI impact the performance of hedge funds? One possible answer is that firms that endorse responsible investment could enhance shareholder value by pressuring firms to improve ESG performance. Alternatively, PRI signatories may constrain their ability to deliver superior investment returns by focusing on a smaller investment opportunity set that comprises stocks with strong ESG performance or that excludes sin stocks. Another important question is: Does hedge funds' endorsement of PRI reflect efforts by money managers to exploit investors' nonpecuniary preference for responsible investment?

The 2021 study *Responsible Hedge Funds* examined what drives the performance of hedge funds managed by PRI signatories.[30] The authors used the Thomson Reuters stock ESG scores to calculate value-weighted portfolio-level ESG scores for investment management firms. They evaluated hedge funds using monthly net-of-fee returns and assets under management data of live and dead hedge funds reported in the Hedge Fund Research (HFR) and Morningstar data sets covering the period 1994–April 2019. Their fund universe had a total of 18,440 hedge funds, of which 3,896 were live funds and 14,544 were dead funds—demonstrating the importance of taking into account survivorship bias. The authors also addressed the issue of incubation bias by dropping all returns data before funds were listed in the data sets.

Their data set included 2,321 PRI signatories. By the end of the sample period in April 2019, there were 174 PRI signatory hedge fund firms managing 489 hedge funds with $316 billion under management, an eleven-fold increase in the hedge fund assets managed by PRI signatories. In addition, during this period, the assets managed by hedge fund firms that endorsed the PRI increased from a modest 3 percent to 30 percent of all hedge fund assets.

The authors calculated firm ESG performance primarily using Thomson Reuters data. The Thomson Reuters ESG ratings measure a company's relative ESG performance, commitment, and effectiveness across 10 main themes: environmental resource use, ecological emissions, environmental product innovation, workforce, human rights, community, product responsibility, management, shareholders, and corporate social responsibility (CSR) strategy. The ratings are derived from more than 400 company-level ESG metrics, which are based on information from annual reports, company websites, nonprofit organization websites, stock exchange filings, CSR reports, and news sources. They complement the Thomson Reuters ESG data with data from MSCI ESG Research (STATS) and Sustainalytics.

The MSCI ESG score is based on strength and concern ratings for seven qualitative issue areas, which include community, corporate governance, diversity, employee relations, environment, human rights and product, as well as concern ratings for six controversial business issue areas, namely, alcohol, gambling, firearms, military, nuclear power, and tobacco.

The Sustainalytics ESG ratings gauge how well companies manage ESG issues related to their businesses and provide an assessment of firms' ability to mitigate risks and capitalize on opportunities. Sustainalytics assesses a company's ESG engagement along four dimensions: (1) preparedness—assessments of company management systems and policies designed to manage material ESG risks, (2) disclosure—assessments of whether company reporting meets international best practice standards and is transparent for most material ESG issues, (3) quantitative performance—assessments of company ESG performance based on quantitative metrics such as carbon intensity, and (4) qualitative performance—assessments of company ESG performance based on the controversial incidents that the company is involved in. The authors found:

- Signatories exhibit better ESG performance than do nonsignatories; the average ESG scores for signatories and nonsignatories were 68.6 and 60.0, respectively. However, 21 percent of signatory ESG scores fell below the median ESG score—a significant number of signatories do not walk the talk.
- ESG scores are highly persistent—ESG performance is a durable characteristic of investment firms.
- Hedge funds managed by investment management firms that endorse the PRI underperformed those managed by other

investment management firms by 2.45 percent per annum (t-stat = 3.93) after adjusting for covariation with the Fung and Hsieh seven factors. The spread in raw returns was 1.44 percent (t-statistic = 2.06).

- The underperformance of signatory hedge funds is substantially stronger in signatories with low ESG scores. Low-ESG signatory hedge funds underperformed low-ESG nonsignatory hedge funds by 7.72 percent per annum (t-statistic = 3.18) after adjusting for risk. In contrast, the difference in risk-adjusted performance between high-ESG signatory and nonsignatory hedge funds was a modest 0.54 percent per annum (t-statistic = 0.74).
- Hedge funds with low ESG exposure underperformed by a risk-adjusted 5.94 percent per year (t-statistic = 3.00) the hedge funds of those with high ESG exposure.
- The results were similar when decomposing the Thomson Reuters ESG score into the component based on environmental and social factors and the component based on corporate governance factors.
- The findings are not driven by smaller hedge funds.
- Signatories who do not walk the talk exhibit greater operational risk.
- While hedge funds that endorsed the PRI underperformed other hedge funds after adjusting for risk, they attracted larger flows and harvested greater fee revenues—signatories attracted an economically and statistically meaningful 16 percent more flows per annum than did nonsignatories.

The authors concluded: "The results suggest that some signatories strategically embrace responsible investment to pander to investor preferences." They added: "The findings suggest that the underperformance of signatory hedge funds cannot be traced to high ESG stocks and, therefore, support the agency [risk, misalignment of interests] view." They also noted: "Low-ESG signatories are more likely to disclose new regulatory actions as well as investment and severe violations on their form ADVs, suggesting that they deviate from expected standards of business conduct or cut corners when it comes to compliance and record keeping." Unfortunately, they also noted: "Investors appear unaware of the agency and operational issues percolating at such signatories. Low-ESG signatories attract as much fund flows as do high-ESG signatories." The bottom line is that some firms appear to strategically endorse responsible investing but don't walk the talk.

Their findings are consistent with *Analyzing Active Fund Managers' Commitment to ESG: Evidence from the United Nations Principles for Responsible Investing,* the 2021 study of all active managers rather than hedge fund managers alone.[31] The authors found "a significant increase in fund flow to signatory funds regardless of their prior fund-level ESG score. However, signatories do not improve fund-level ESG score while exhibiting a decrease in return." The decrease in returns is not related to decreasing economies of scale. They also found that only quant-driven and institution-only funds improve their ESG scores post signing: "Overall, only a small number of funds improve ESG while many others use the PRI status to attract capital without making notable changes to ESG." And finally, they shockingly found that "signatories vote less on environmental issues and their stock holdings experience increased environment related controversies." It is a shock, they added, because "environmental controversies have been documented to be tail risks that have significant negative implications to stock prices."

The 2022 study *Do Responsible Investors Invest Responsibly?* found similar results for U.S. domiciled institutional funds: U.S. institutions that publicly commit to responsible investing do not exhibit better ESG scores.[32] However, non-U.S. institutions that publicly commit to PRI principles do exhibit higher ESG scores. Consistent with other research, they also found "weak evidence of lower equity portfolio returns when comparing them to non-PRI signatories." However, they also found "evidence that negative screening, integration, and engagement lower portfolio risk."

Unfortunately, the evidence demonstrates that at least a significant portion of funds use PRI as a marketing ploy and a way for companies to get free money. And for hedge funds, there is evidence that responsible investing has negatively impacted returns. The same is true for institutional funds in general, though the evidence of a negative impact on returns is weaker.

1.2.6 Investors Benefit from Hedge Funds Identifying Undervalued Stocks

The role of hedge funds in stock price formation was extensively examined in the 2018 study *Hedge Funds and Stock Price Formation.*[33]

The authors, Charles Cao, Yong Chen, William Goetzmann, and Bing Liang, focused on determining whether hedge funds, as a group, exploit and correct price inefficiencies in the stock market. Using the long-position data (long positions target what the buyer perceives to be undervalued stocks), they studied the role of hedge funds in the stock price formation process.

Note that because the SEC does not require institutions to disclose their short positions (which seek overvalued stocks), their analysis focused on the long positions and positive-alpha stocks. Their dataset consisted of stock holdings of 1,517 hedge fund management companies and covered the period 1981–2015. The SEC requires hedge fund companies with more than $100 million in assets under management to file quarterly disclosures of equity holdings. Thus, portfolios are rebalanced quarterly, which has the benefit of controlling trading costs. Following is a summary of their findings:

- By 2015, hedge funds controlled 16.4 percent of shares held by all institutions, while mutual funds and banks controlled 39.2 percent and 14.4 percent, respectively.
- Stocks with high hedge fund ownership have lower dividend yields, younger age, and a lower percentage of the S&P 500 Index membership in comparison with the entire sample.
- Hedge funds tend to hold undervalued stocks—stocks that go on to outperform, generating alpha relative to the *Fama-French four-factor* (beta, size, value, and momentum) model—and thus can identify mispricings.
- Undervalued stocks, relative to stocks with insignificant alphas, are associated with higher hedge fund ownership (statistically significant at the 1 percent confidence level).
- Hedge fund ownership is not significantly related to negative-alpha stocks.
- Both hedge fund ownership and trades are positively related to the degree of mispricing—hedge funds increase their purchases with the degree of underpricing, but this is not the case for non-hedge funds. A portfolio of positive-alpha stocks with high hedge fund ownership realized a risk-adjusted return of 0.40 percent (t-statistic = 3.36) per month, about 4.8 percent per year, significantly

outperforming a counterpart portfolio of positive-alpha stocks with low hedge fund ownership (0.02 percent per month; t-statistic = 0.16). Notably, the outperformers were not less liquid stocks—trading costs were manageable and easily implementable. The alpha exceeded conventional estimates of trading cost. These are long-only portfolios, avoiding the high costs often associated with shorting. Although the high ownership portfolio has higher return volatility, it exhibited a higher Sharpe ratio.

- A portfolio with large hedge fund trades significantly outperformed a portfolio with small trades. For example, the large trade portfolio shows an alpha of 0.36 percent (t-statistic = 3.21) per month, significantly higher than the alpha of 0.04 percent (t-statistic = 0.32) per month for the small trade portfolio. In contrast, there was little difference between the portfolios formed by the trades of non-hedge fund institutions.

- Undervalued stocks with higher hedge fund ownership and trades in one quarter were more likely to have mispricing corrected in the next quarter, suggesting that hedge funds help reduce mispricing. However, price correction does not occur instantaneously.

- For non-hedge fund institutional investors, including banks, insurance companies, and mutual funds, their stock ownership, on average, was neither related to stock underpricing nor predictive for stock returns.

- There was a significant relation between lagged idiosyncratic volatility and hedge fund ownership (but not non-hedge fund ownership). In contrast, there was no significant relationship between the trades of non-hedge fund institutions and idiosyncratic volatility. This finding is consistent with the view that hedge funds bear arbitrage costs when exploiting price inefficiencies.

The above findings led the authors to conclude that hedge funds play an essential role in the security price formation process and help to make the market more efficient.

Let's now examine if investors benefit from the alpha generated from identifying undervalued stocks. We start with the 4.8 percent alpha reported by the authors and adjust it down to 4.0 percent after accounting for trading costs and market impact. Historically, the U.S. stock market has returned about 10 percent. A 4 percent alpha would mean that hedge funds would produce a gross return of 14 percent. We can now apply the typical 2/20 fee to that gross return. Subtracting the 2 percent expense ratio reduces the gross return to 12 percent. Subtracting the incentive fee of 20 percent would reduce that to 9.6 percent. Therefore, hedge fund investors may be disappointed by their net-of-fee performance despite the hedge fund managers having a meaningful edge in selecting mispriced stocks.

1.3 KEY TAKEAWAYS

Following are the takeaways from this chapter:

- The hedge fund industry has a long and rich history of innovation. Its strategies have evolved from the early long-short technical signals of Alfred Winslow Jones in 1949 to modern systematic trading and machine learning approaches.
- Hedge fund investing is often misunderstood. We debunked several myths: hedge fund investing and selection of top performers are easy, hedge funds hedge, active and socially responsible hedge funds outperform, and investors benefit from hedge funds identifying undervalued stocks.

CHAPTER 2

Hedge Fund Research and Data

"An investigator starts research in a new field with faith, a foggy idea, and a few wild experiments. Eventually the interplay of negative and positive results guides the work. By the time the research is completed, he or she knows how it should have been started and conducted."—Donald Cram.[a]

"I start early and I stay late, day after day, year after year. It took me 17 years and 114 days to become an overnight success."—Lionel Messi.[b]

We continue our journey into the challenging field of hedge fund research. This chapter poses essential research questions about hedge funds and discusses hedge fund databases and their inherent biases that create issues for empirical research.

2.1 RIGOROUS AND PRACTICAL HEDGE FUND RESEARCH

Having debunked several popular myths about hedge funds, over the next several chapters we address critical questions about hedge fund investing that are relevant for investors.

- What are the drivers of hedge fund performance?

[a]Donald Cram was a Nobel Laureate in chemistry, 1987.
[b]Lionel Messi is regarded as one of the greatest soccer players of all time.

DOI: 10.1201/9781003175209-2

- Can we predict future performance?
- Which hedge funds should an investor choose?
- What is the best way to combine hedge funds into a single portfolio?

What is the process for answering challenging questions about hedge fund investing? Although hedge fund research is a fruitful area for both academics and practitioners, they tend to have very different perspectives. We can think of academics as artists and practitioners as carpenters. Academics crave beauty and look at the world through the lenses of an artist with little regard for utility. By contrast, practitioners are like carpenters. They appreciate durability and usefulness. Artists and carpenters are inspired differently. An artist may spend hours contemplating *The Night Cafe,* the masterpiece by Vincent van Gogh, analyzing how Van Gogh expressed the terrible passions of humanity through red and green. By contrast, a carpenter may quickly lose interest in looking at the three sleeping drunks and instead choose to watch a *Do It Yourself* video to obtain some useful nuggets. Each perspective is valuable but incomplete. Just as you need to combine both perspectives if you want to build a beautiful and durable restaurant, you need to appreciate and blend both views if you are going to construct a robust hedge fund portfolio.

Over the years, academics have provided valuable insights into certain aspects of portfolio management, particularly regarding fund evaluation and portfolio construction. Yet, there remains a lack of a widely accepted, robust, and flexible methodology that can evaluate whether those insights can benefit a specific institutional investor once implemented with real-world constraints. It takes an artist and a carpenter working together to accomplish that.

2.1.1 Challenges

Several important challenges need to be carefully considered. First, investors have their unique objectives and constraints that vary depending on the type of institution. For example, a family office or an asset management firm might seek to maximize a Sharpe ratio. A university endowment might attempt to target returns that exceed the university's spending rate over a market cycle. A pension fund might pursue maximization of risk-adjusted returns within an asset-liability framework.

Moreover, sophisticated investors often impose rigorous filtering criteria such as the length of a track record and level of assets under management (AUM). Unfortunately, most academic studies either ignore these critical criteria, or they selectively incorporate some of them to account for certain biases (such as small fund bias or incubation bias). While accounting for biases is essential, an institutional investor ultimately wants to know whether they will be able to benefit from a portfolio management technique given its own set of preferences and constraints.

Second, testing methodologies need to be relevant and implementable. Most academic papers compare portfolios that include hundreds of funds. Thus, their findings may have little value for an investor who plans to hire three to five hedge funds. Moreover, the results of many hedge fund studies are not implementable because they ignore delays in hedge fund reporting. While popular testing methodologies often come from research on equities and mutual funds with their daily returns available without delay, hedge fund databases rely on self-reporting by hedge fund managers and provide monthly returns with a delay of about one month. This issue introduces a look-ahead bias (a decision in an empirical study is made using data that was not readily available at the time of the decision) in most hedge fund studies.

It's not uncommon for academic research to become disconnected from reality. One reason is that academic research exists within an ecosystem of peer-reviewed journals. Once an approach is established, it gets added to the toolbox of standard techniques. If someone discovers a problem with the original method and writes a paper about it, the paper will be reviewed by an expert in the field who has probably used the initial approach. As Hans Christian Andersen showed in his famous tale, it takes a lot of courage to publicly declare that the emperor has no clothes. While difficult, it does happen occasionally.

Ivo Welch, the author of the 2013 study *A Critique of Recent Quantitative and Deep-structure Modeling in Capital Structure Research and Beyond,* demonstrated courage as he boldly criticized the deep-structure modeling approach widely regarded as the leading state-of-the-art method in theoretical corporate finance.[1] Corporate finance is the study of the behavior of firms. Structural models are typically very complex dynamic models with many assumptions that allow the decision-maker (such as a firm's chief financial officer) to optimize decisions (such as the choice of financial leverage). Such models are typically evaluated based on the underlying assumptions and provide estimates that match data moments. However, Welch showed that the models often omit plausible forces not based on evidence but on authors' beliefs and that tests largely ignored important econometric issues, such as selection and survivorship biases, discussed in detail in this chapter.

For example, as discussed earlier, hedge fund performance persistence tests must account for delays in hedge fund reporting. While adjusting for hedge fund reporting delays is still uncommon because of a lack of familiarity with reporting practices, academic research generally attempts to carefully consider data availability. A good example is the 1992 study *The Cross-section of Expected Stock Returns* in which the authors, Eugene Fama and Kenneth French, relied on a six-month lag to sufficiently account for the delay in accounting reporting for measuring the book-to-market ratio, a vital component of the value factor.[2]

2.2 HEDGE FUND DATA: GARBAGE IN, GARBAGE OUT?

Hedge fund researchers use empirical hedge fund data to answer portfolio management questions such as the existence of skill among hedge fund managers, persistence in their performance, and identifying a portfolio construction edge. Thus, it is critical to understand hedge fund data and potential issues that arise in empirical research.

2.2.1 Public Databases and Biases

One issue is that researchers can potentially draw different conclusions depending on their choice of hedge fund database and a set of approaches to mitigate the biases in the data. The task of creating a reliable bias-free database is challenging for several reasons. First, since hedge fund reporting is voluntary for strategic advertising and asset raising reasons, as discussed in the 2014 study *The Strategic Listing Decisions of Hedge Funds* by Philippe Jorion and Christopher Schwarz, hedge fund managers have an incentive to report only good performance.[3] They may delay or even misreport poor performance, as discussed in detail in the 2009 paper *Do Hedge Fund Managers Misreport Returns? Evidence from the Pooled Distribution* by Nicolas Bollen and Veronika Pool.[4] Second, hedge fund data is subject to biases that need to be carefully examined and mitigated, as discussed later in this chapter.

In the 2021 paper *Hedge Fund Performance: Are Stylized Facts Sensitive to Which Database One Uses?*, the authors, Juha Joenvaara, Mikko Kauppila, Robert Kosowski, and Pekka Tolonen, performed a comprehensive study of five commercial databases (BarclayHedge, EurekaHedge, Hedge Fund Research (HFR), Lipper TASS, and Morningstar) commonly used for academic research and two commercial databases (eVestment and Preqin) that are rarely used by researchers.[5]

Among the problems with hedge fund databases is that most report "duplicate" share classes of the same fund. The share classes may have different investment terms, such as onshore/offshore status, currency class, or fees. Typically, researchers will eliminate duplicate share classes and select a single share class based on the length of the track record, the amount of assets under management, or the expense ratio. Instead of choosing a representative share class, the authors recommended following mutual fund literature and aggregating the fund-level information across all duplicates and databases. They showed that their approach overcomes the issue of incomplete fund data in individual databases and results in broader coverage of the hedge fund universe available for researchers.

While the authors recommended aggregating all seven databases, they found that their results regarding the average performance of hedge funds and the performance persistence were very similar when utilizing two individual "research-quality" databases: HFR and BaclayHedge. Their analysis of the seven databases can be summarized as follows:

- The HFR database, managed by Hedge Fund Research, Inc, is an excellent choice for researchers with consistently high coverage of return, AUM, and fund characteristic information. The dataset is free of survivorship bias since 1994. Its only problem is poor commodity trading advisors (CTA) coverage during early periods, as reported in the 2002 paper *Hedge Fund Benchmarks: Information Content and Biases* by William Fung and David Hsieh.[6]

- The BarclayHedge, managed by BarclayHedge, a division of Backstop Solutions, has a minor problem of missing share restriction variables during early periods. However, it has the most extensive CTA coverage at 2,944 funds, whereas the other databases have between 650 and 1,449 funds. Thus, BarclayHedge is the best database for CTA research.

- Lipper TASS, managed by Lipper, used to be a high-quality database, but the acquisition of Trading Advisor Selection System, TASS, in 2005 resulted in spurious survivorship bias reported in the 2010 study *Hidden Survivorship Bias in Hedge Fund Returns* by Rajesh Aggarwal and Philippe Jorion, and incorrect values in the fields used for mitigating the backfill bias, as reported in the 2009 paper *Measurement Biases in Hedge Fund Performance Data: An Update* by William Fung and David Hsieh.[7,8]

- Morningstar, managed by Morningstar, combines CISDM, formerly CTA-heavy MAR database, Altvest, MSCI/Barra, and gathers data from the quarterly SEC holdings report by funds of hedge funds. While it used to be a high-quality database, it often fails to report AUM information during later periods.

- EurekaHedge, owned by Mizuho Corporate Bank, is relatively new. It was created in 2001 and has high quality coverage of European and Asian funds.

- The commercial databases (eVestment and Preqin) are of poor quality.

We now turn to discussing hedge fund data biases that need to be carefully examined and mitigated by researchers.

Common biases in hedge fund returns include:

- **Selection bias.** The selection bias emerges from the voluntary nature of reporting to hedge fund databases. There is mixed evidence regarding the severity of the selection bias. The authors of the 2013 study *Out of the Dark: Hedge Fund Reporting Biases and Commercial Databases* examined regulatory filings of fund-of-funds registered with the SEC that include quarterly holdings and estimated the impact of selection bias to be approximately 3–5 percent per annum by directly comparing the performance from the SEC filings and the performance in commercial databases.[9] By contrast, the 2013 paper *Exploring Unchartered Territories of the Hedge Fund Industry: Empirical Characteristics of Mega Hedge Fund Returns* by Daniel Edelman, William Fung, and David Hsieh showed that the selection bias was likely insignificant.[10]

  As discussed in the 2014 study *The Strategic Listing Decisions of Hedge Funds* by Philippe Jorion and Christopher Schwarz, hedge fund managers report to databases for strategic advertising.[11] Two types of funds that choose not to report their performance to public databases: poorly performing funds that cannot attract new investors based on their track record and successful funds that do not rely on commercial databases to raise assets. In the 1999 paper *The Performance of Hedge Funds: Risk, Return, and Incentives,* the authors documented that some successful hedge funds stop reporting because of the diminishing returns to their arbitrage strategies.[12] Edelman, Fung, and Hsieh showed that the two sources of non-reporting bias cancel each other out, and the performance of mega-funds that don't report to commercial databases was similar to those that do.

Moreover, Joenvaara, Kauppila, Kosowski, and Tolonen showed that high-quality databases such as HFR or BarclayHedge result in similar inferences regarding average hedge fund performance and performance persistence as an aggregate of seven databases.[13] Thus, some evidence suggests that selection bias may not materially alter research findings when high-quality databases are used for empirical analysis. However, the aforementioned 2021 study *The Hedge Fund Industry is Bigger (and has Performed Better) Than You Think* by Daniel Barth, Juha Joenvaara, Mikko Kauppila, and Russ Wermers discovered strong evidence of selection bias.[14] The authors compared the performance of hedge funds from commercial databases (publicly reporting funds) and the performance from the regulatory data for U.S. funds (non-publicly reporting funds) that do not report to any public database. They found that non-publicly reporting funds delivered superior risk-adjusted performance relative to the publicly reporting funds.

- **Survivorship bias.** The issue of survivorship bias is not unique to finance. One of the most prominent examples is the story of the famous statistician Abraham Wald, a member of the Statistical Research Group at Columbia University that helped the U.S. military to minimize bomber losses to enemy fire during World War Two. As described in the 1980 paper *The Statistical Research Group, 1942–1945* by Allen Wallis, the U.S. military was considering reinforcing the areas that were the most damaged areas of the planes that made it back, a conclusion subject to the survivorship bias.[15] By contrast, the Statistical Research Group came up with a justification for adding armor to the areas with minor damage because they were more critical for survival.

  The 2000 study *Hedge Funds: the Living and the Dead* by Bing Liang highlighted the role of survivorship bias in hedge fund databases and found that it exceeded 2 percent per annum.[16] The 1999 paper *Offshore Hedge Funds: Survival and Performance 1989–1995* by Stephen Brown, William Goetzmann, and Roger Ibbotson estimated survivorship bias to be close to 3 percent.[17]

A simple way to illustrate the impact of survivorship bias on inferences is to consider a hypothetical situation where a hedge fund that is down 50 percent is expected to lose an additional 50 percent and go out of business with a 75 percent probability over the next year or to make 10 percent with a 25 percent probability. Suppose an average hedge fund makes 5 percent, on average. In that case, considering only survived funds will show that investing in a fund that is down 50 percent is an excellent investment opportunity because the fund is "expected" to be up 10 percent, or twice as much as an average fund. However, including defunct funds would change the expected return to $-50\% \times 0.75 + 10\% \times 0.25 = -35\%$, a very poor performance number.

The 1992 paper *Survivorship Bias in Performance Studies* by Stephen Brown, William Goetzmann, Roger Ibbotson, and Stephen Ross showed that not properly accounting for survivorship bias may lead to the appearance of predictability in mutual fund returns.[18] A standard approach to mitigating survivorship bias is to include the "graveyard" database of defunct funds.

- **Backfill/incubation bias.** The backfill and incubation biases arise due to the voluntary nature of self-reporting. Typically funds go through an incubation period during which they build a track record using proprietary capital. Fund managers start reporting to a hedge fund database to raise capital from outside investors only if the track record is attractive. Unfortunately, they are often allowed to "backfill" the returns generated prior to their inclusion in the database. Since funds with poor performance are unlikely to report returns to the database, incubation/backfill bias results. The 2005 study *Hedge Funds: Risk and Return* by Burton Malkiel and Atanu Saha used the Lipper TASS database and estimated the backfill bias to be roughly 5 percent per annum.[19]

  There are three common approaches to mitigating the backfill bias:

  1. *Truncating a fixed number of returns (typically 12–30 monthly returns).* In their aforementioned 2007 study *Do Hedge Funds*

Deliver Alpha? A Bayesian and Bootstrap Analysis, Robert Kosowski, Narayan Naik, and Melvyn Teo recommended removing the first 12 monthly returns.[20] This approach was the most common approach in academic literature until very recently when several studies showed that it was insufficient.

For example, the 2014 paper *Fooling Some of the People All of the Time: The Inefficient Performance and Persistence of Commodity Trading Advisors* by Geetesh Bhardwaj, Gary Gorton, and Geert Rouwenhorst showed that the standard adjustment of removing 24 months of returns was inadequate for CTAs because the remaining bias was still more than 1 percent per annum in a value-weighted index and almost 3 percent in an equally-weighted index.[21]

Similarly, in the 2017 study *The Fix is In: Properly Backing Out Backfill Bias,* Philippe Jorion and Christopher Schwarz showed that the truncated approach based on 24 months retained approximately 70 percent of backfilled returns.[22] The aforementioned 2009 paper *Measurement Biases in Hedge Fund Performance Data: An Update* by William Fung and David Hsieh showed that backfill periods can sometimes cover 10 years.[23]

2. *AUM-based.* The 2010 study *Luck Versus Skill in the Cross-section of Mutual Fund Returns* by Eugene Fama and Kenneth French suggested limiting the dataset to those funds that managed a minimum acceptable AUM normalized to the end-of-period values using CPI as a proxy of inflation (e.g., least US $10 million in AUM normalized to December 2014 values).[24] Once a fund reaches the AUM minimum, it is included in all subsequent tests to avoid creating selection bias. Unfortunately, many hedge funds initially reported only net returns for an extended period of time before their initial inclusion of AUM data. The AUM-based approach would eliminate large portions of valuable data for such funds.
3. *Using the first reported date field.* Hedge fund databases often include a field that provides information about each hedge fund's first reported date. An intuitive way to mitigate the backfill bias is to remove all returns before the first reported date. As shown in the aforementioned 2014 paper by Bhardwaj et al., the 2017 study *Just a One Trick Pony?*

An Analysis of CTA Risk and Return by Jason Foran, Mark Hutchinson, David McCarthy, and John O'Brien, and the 2017 paper *The Fix is In: Properly Backing Out Backfill Bias* by Philippe Jorion and Christopher Schwarz, this approach provides the best adjustment for the backfill bias.[25,26]

Its only weakness is that most hedge fund databases either didn't have the first reported date field at launch or discontinued it. For example, the BarclayHedge database started using that field in 2002, and all funds that reported to the database before 2002 have the first reported date set to December 2002. The Lipper TASS database stopped providing that field in March 2011.

The two 2017 studies just cited suggested two algorithms for inferring the first reported date. Jorion and Schwarz used the fact that fund IDs are typically assigned in chronological order when they are added to the database. Since the BarclayHedge requires that funds report to the database to be considered for inclusion in the Barclay CTA index, Foran, Hutchinson, McCarthy, and O'Brien estimated the first reported date of a fund before 2002 by the first date that the fund is included in the index.

- **Liquidation bias.** As discussed in detail in the aforementioned 1999 paper *The Performance of Hedge Funds: Risk, Return, and Incentives*, some hedge fund managers strategically choose not to report the last (and likely) poor performance numbers of defunct funds to databases because that would only hurt their ability to raise assets for their remaining funds.[27] Thus, defunct funds could lose substantial value following the last reported date. The authors worked with HFR to poll each of the defunct funds in the HFR database to recover all returns from the last reported date to the final termination date. This comprehensive study showed that the liquidation bias, the loss beyond the information already contained in the database, was approximately 0.7 percent.

- **Graveyard bias.** The graveyard bias, reported in the 2014 paper *Fooling Some of the People All of the Time: The Inefficient Performance and Persistence of Commodity Trading Advisors* by Geetesh Bhardwaj, Gary Gorton, and Geert Rouwenhorst, is a type of survivorship bias driven by hedge fund managers requesting database providers to remove complete track records of their defunct funds from the graveyard databases.[28] Thus, instead of the track records transitioning from active hedge fund databases to graveyard databases, which researchers typically use to mitigate survivorship bias, they may completely disappear per hedge fund managers' requests. In addition to introducing survivorship bias, this practice may lead to differences in versions of data available depending on the vintage of the database used for analysis. Bhardwaj, Gorton, and Rouwenhorst confirmed with the Lipper TASS that the practice of removing funds exists because it is consistent with the principle of entirely voluntary reporting. The graveyard bias impacted almost 20 percent of CTAs between April 2008 and September 2012 vintages of the Lipper TASS database and had resulted in an approximately 1.88 percent upward impact on performance. Fortunately, not all databases are subject to graveyard bias. We confirmed that the BarclayHedge database doesn't remove track records.

- **Data revision bias.** The comprehensive 2015 study *Change You Can Believe In? Hedge Fund Data Revisions* by Andrew Patton, Tarun Ramadorai, and Michael Streatfield considered vintages of the five databases (i.e., Lipper TASS, HFR, CISDM, Morningstar, and BarclayHedge) recorded at different points between 2007 and 2011.[29] The authors showed that about 45 percent of hedge funds had revised their previous returns, and over 20 percent of funds had revised a monthly return by at least 1 percent, which is substantial given the average monthly return in the study of 0.62 percent. This behavior is not driven by data entry issues accounting for less than 2 percent of all revisions. It seems to be strategic since most revisions are negative, and about half of all revisions relate to returns that are more than 12 months old. This finding suggests that hedge fund managers attempt to advertise strategically by initially reporting inflated performance to attract clients and then adjusting it. One approach to adjusting for the data revision bias is to compare the performance of the same funds across different vintages of databases. For example, the BarclayHedge database provides monthly vintages.

- **Look-ahead bias.** Hedge fund databases provide returns with approximately one month delay. This delay is usually ignored in academic studies such as the 2004 Journal of Empirical Finance paper *Analysis of Hedge Fund Performance*, the aforementioned 2007 Journal of Financial Economics paper *Do Hedge Funds Deliver Alpha? A Bayesian and Bootstrap Analysis,* and the 2010 Journal of Finance paper *Do Hot Hands Exist Among Hedge Fund Managers? An Empirical Evaluation.*[30,31,32]

These studies form end-of-year portfolios using December returns that are not available to investors until the end of January of the following year. This procedure introduces a look-ahead bias and creates a significant barrier to implementing the results of most studies since investment recommendations are based on information that is not available at the time of investment decision. The most notable example of adjusting for data availability in academic research is the accounting book value in the definition of book-to-market used in the aforementioned 1992 paper *The Cross-section of Expected Stock Returns.* As we discussed, the authors suggested utilizing a 6-month lag which is sufficient to account for delay in accounting reporting.[33]

The 2016 study *A Simulation-based Methodology for Evaluating Hedge Fund Investments* by Marat Molyboga and Christophe L'Ahelec and the 2017 paper *Assessing Hedge Fund Performance with Institutional Constraints: Evidence from CTA Funds* by Marat Molyboga, Seungho Baek, and John Bilson recommended using one month delay for CTAs, a subset of hedge funds in managed futures.[34,35]

2.2.2 Other Frictions and Considerations

Most investors allocate to hedge fund managers through funds or managed accounts. Let us consider a simple example: an investor who wants to allocate $200 million equally to two hedge fund: $100 million to program A offered by manager A and $100 million to program B offered by manager B. Fund investing involves purchasing shares of fund A, subject to the fund liquidity terms such as subscription and redemption notices, lock-ups and gates, and shares of fund B that may have a different set of liquidity terms. The positions of each fund are owned by the hedge fund manager who manages the fund, places trades, and posts margins with clearing firms. By contrast, a managed account is an investment account owned by an investor and managed by the hedge fund manager who places the trades on behalf of the investor. However, the investor is responsible for posting the margin and, thus, can notionally fund the account by depositing only a portion of the nominal funding and increasing the deposit amount in response to margin calls. In this example, the investor may choose to keep the same nominal allocation of \$100 million to program A and \$100 million to program B but only deposit \$20 million in each account. In addition, the liquidity terms of managed accounts are typically less strict than those of funds. For example, most CTAs require notice periods between five and 30 days for their funds' subscriptions and redemptions and offer daily liquidity for managed accounts.

We can illustrate the implications of the difference in liquidity terms by considering a hypothetical *rebalancing* decision made at the end of February after program A makes \$100 million increasing the fund allocation in program A to \$200 million while the fund allocation in program B remains at \$100 million. Since the investor prefers allocating equally, the rebalance decision involves redeeming \$50 million from program A and allocating an additional \$50 million to program B to obtain an equal allocation of \$150 million to each program. While this rebalancing decision is easy to implement with managed accounts with daily liquidity, it is more challenging with funds because of rebalancing frictions.

The timeline of the rebalancing process once the decision is made at the end of February:

1. On March 1st, the redemption order for \$50 million is sent to manager A.

2. On March 31st, the redemption becomes effective, and the exposure to program A is reduced by \$50 million. It does not mean that the exposure will be equal to \$150 million because of the March return of program A. If, for example, fund A makes an additional \$20 million, the post-redemption exposure to fund A is \$200 million + \$20 million −\$50 million = \$170 million. However, if the fund loses \$20 million in March, the post-redemption exposure would be \$200 million −\$20 million −\$50 million = \$130 million.

3. On April 1st, the redemption amount of \$50 million is out of the market, but the investor has to wait until the end of the month to receive it.

4. On April 30th, the investor receives the redemption amount of \$50 million.

5. On May 1st, the investor wires the money to manager B to subscribe at the end of May.

6. On May 31st, the subscription is finalized.

7. On June 1st, the exposure in program B increases by \$50 million. Once again, the total post-subscription investment is not equal to the target value of \$150 million because of the gains and losses in March, April, and May.

The rebalancing process is not only cumbersome but also costly for several reasons. First, the \$50 million amount is uninvested between April 1st and May 31st. Therefore, the investor is under-exposed to hedge funds for two months purely due to the fund rebalancing frictions. Second, the investor cannot allocate equally between the hedge funds because of the delays. Finally, if an investor relies on a more sophisticated allocation approach, the impact of frictions can be greater. Therefore, investors should either allocate via managed accounts or carefully account for rebalancing frictions in their investment process.

2.3 KEY TAKEAWAYS

Following are the key takeaways from this chapter:

- Empirical hedge fund research is performed by academics and practitioners. Each perspective is valuable but incomplete, making it vital to appreciate and blend both views in order to build robust hedge fund portfolios.

- Hedge fund researchers can use one or several databases. However, the quality of the databases varies significantly. The HFR database is an excellent choice for researchers as long as they don't specialize in CTAs. BarclayHedge is the best database for CTA research.

- Empirical data should be evaluated and adjusted for the following biases: selection, survivorship, backfill/incubation, liquidation, graveyard, data revision, and look-ahead.

- Investors should either allocate via managed accounts or carefully account for rebalancing frictions in their investment process.

- We posed four important research questions:

    1. What are the drivers of hedge fund performance?
    2. Can we predict future performance?
    3. Which hedge funds should an investor choose?
    4. What is the best way to combine hedge funds into a single portfolio?

  The first one will be answered in detail in Chapter 3. The second and third questions will be addressed in Chapter 4. The fourth question will be covered in Chapters 5 and 6.

CHAPTER 3

Manager Selection and Hedge Fund Factors

"Know what you own, and know why you own it."—Peter Lynch.[a]

"In the National Football League you get one first-round draft pick if you're lucky. You couldn't really outwork anybody else. In college I could recruit ten players with first-round talent every year."—Nick Saban.[b]

After learning about the exacting aspects of empirical hedge fund research, we turn to one of the most challenging topics in portfolio management. This chapter provides a thorough overview of the essential aspects of manager selection including quantitative and qualitative analysis. It demonstrates that a gap exists between standard academic methodologies and industry needs and then proposes a robust and implementable framework for evaluating manager selection techniques. It also describes factors and factor selection techniques for performance evaluation of hedge funds.

3.1 ACADEMIC RESEARCH AND FRAMEWORK FOR MANAGER SELECTION

The topic of manager selection includes several key areas: factor models, factor selection, performance evaluation, and performance persistence. Performance persistence tests in hedge funds are based on the standard

[a]Peter Lynch is a famous American investor who managed the best-performing mutual fund in the world.

[b]Nick Saban is regarded as one of the greatest coaches in college football history.

DOI: 10.1201/9781003175209-3

approaches developed for testing momentum, or short-term persistence in relative performance, in various asset classes and mutual funds.

Momentum has been documented in U.S. equities, international equities, industries, equity indices, foreign exchange markets, global bond markets, and commodities. The 2013 paper *Value and Momentum Everywhere* by Clifford Asness, Tobias Moskowitz, and Lasse Pedersen presented results of a comprehensive study of cross-sectional momentum and value strategies across several asset classes including individual stocks, stock indices, currencies, commodities, and bonds.[1] The authors found significant momentum in every asset class considered in the study.

The 1993 Journal of Finance paper *Hot Hands in Mutual Funds: Short-run Persistence of Relative Performance, 1974–1988* by Darryll Hendricks, Jayendu Patel, and Richard Zeckhauser tested for momentum in mutual fund returns.[2] The authors found persistence in relative performance of mutual funds with the difference in the risk-adjusted performance of the top and bottom octile portfolios of 6–8 percent per year. Similarly, the 1997 study *On Persistence in Mutual Fund Performance* by Mark Carhart used a decile methodology to evaluate persistence in mutual fund performance and found strong persistence in performance of the worst performing managers and no evidence of skilled or informed mutual fund portfolio managers who consistently provide better risk-adjusted returns.[3]

3.1.1 Hedge Fund Performance Persistence

The techniques used to test for momentum in various asset classes and mutual funds are often relevant to institutional investors, who can relatively easily build large long-short portfolios of "winners minus losers" and rebalance them monthly, although these investors still need to deal with practical implementation issues of transaction costs and market impact. Very similar "portfolio sorting" techniques are used to evaluate persistence in performance of hedge funds.

For example, the 2004 paper *Analysis of Hedge Fund Performance* by Daniel Capocci and Georges Hubner used a decile methodology to discover the lack of persistence among the top and bottom decile funds as well as little persistence among middle decile funds.[4] Two 2020 studies *Multi-period Performance Persistence Analysis of Hedge Funds* and *On Taking the Alternative Route: The Risks, Rewards, and Performance Persistence of Hedge Funds* by Vikas Agarwal and Narayan Naik documented a meaningful quarterly persistence of hedge fund returns primarily driven by the worst performing funds.[5,6] The 2007 study *Do Hedge Funds Deliver Alpha? A Bayesian and Bootstrap Analysis* by Robert Kosowski, Narayan Naik, and Melvyn Teo applied estimates of alphas calculated using the Bayesian methodology introduced in the 2002 paper *Mutual Fund Performance and Seemingly Unrelated Assets,* to demonstrate performance persistence over a one-year horizon.[7,8]

The 2010 study *Do Hot Hands Exist Among Hedge Fund Managers? An Empirical Evaluation* by Ravi Jagannathan, Alexey Malakhov, and Dmitry Novikov used weighted least squared and *General Method of Moments* (GMM) approaches to find significant performance persistence among the top performing hedge funds and little evidence of persistence among the bottom performing funds.[9] The authors ranked funds using the t-statistic of alpha and reported superior performance of portfolios of all funds in the top decile and tercile of all funds—a discovery of particular importance for institutional investors who attempt to identify top-performing hedge fund managers.

Unfortunately, as discussed in the 2016 study *A Simulation-Based Methodology for Evaluating Hedge Fund Investments* by Marat Molyboga and Christophe L'Ahelec, and the 2017 paper *Assessing Hedge Fund Performance with Institutional Constraints: Evidence from CTA Funds* by Marat Molyboga, Seungho Baek, and John Bilson, standard portfolio sorting techniques cannot be implemented by prudent institutional investors because they are not consistent with investment practices and real-world constraints.[10,11] There are several important aspects that must be carefully considered when trying to assess practical benefits of a manager selection approach:

1. Investors have objectives that vary substantially, depending on the type of institution they represent. For example, an asset management firm might seek to maximize a Sharpe ratio, while a university endowment might attempt to achieve returns that exceed the universityâĂŹs spending rate over a market cycle, or a pension fund might try to maximize risk-adjusted return within an asset-liability framework.

2. Sophisticated investors often utilize rigorous filtering criteria such as length of track record or level of AUM. Most academic studies either completely ignore these selection criteria or selectively incorporate them with the purpose of accounting for certain biases such as the small fund bias or incubation bias. For example, the aforementioned 2004 paper *Analysis of Hedge Fund Performance*

and the 2010 study *Do Hot Hands Exist Among Hedge Fund Managers? An Empirical Evaluation* didn't impose a minimum AUM requirement and required minimum track record lengths of 12 months and 36 months, respectively.[12,13] While accounting for biases is important to ensure validity of empirical results, institutional investors also need to evaluate investment decisions given their own sets of preferences and constraints.

3. Most academic papers often compare portfolios that include hundreds of funds. For example, the 2010 study *Do Hot Hands Exist Among Hedge Fund Managers? An Empirical Evaluation* considered tercile portfolios with 252 funds and decile portfolios with 77 funds.[14] Unfortunately, the findings of this study may not be directly relevant for the majority of investors who allocate to a much smaller number of hedge funds. Such investors would be interested in evaluating the impact of their manager selection and portfolio construction decisions on the distribution of potential outcomes. Generating out-of-sample results for portfolios with a smaller number of managers within a simulation framework can provide this information.

4. While investors care about the marginal impact of hedge fund investments on their existing portfolio, this impact is often ignored in traditional analyses.

5. Hedge fund databases provide delayed monthly returns. As discussed in detail in Section 2.2.1, this delay is usually ignored in academic papers, which introduces a look-ahead bias, creating a significant barrier to implementing the results of most studies since investment recommendations are based on information that is not available at the time of investment decision. The aforementioned 2016 paper *A Simulation-Based Methodology for Evaluating Hedge Fund Investments* and the 2017 paper *Assessing Hedge Fund Performance with Institutional Constraints: Evidence from CTA Funds* recommended using a one-month delay for CTAs, a subset of hedge funds that invest in highly liquid instruments.[15,16] A delay of three months should be sufficient for most hedge fund strategies.

The failure to account for these common industry constraints may limit the applicability of academic research for investors.

3.1.2 General Framework of Molyboga, Bilson, and Baek

The 2016 paper *A Simulation-Based Methodology for Evaluating Hedge Fund Investments* by Marat Molyboga and Christophe L'Ahelec, and the 2017 paper *Assessing Hedge Fund Performance with Institutional Constraints: Evidence from CTA Funds* by Marat Molyboga, Seungho Baek, and John Bilson, attempted to close the gap between the academia and the industry by introducing a robust and flexible methodology capable of evaluating whether a fund selection or a portfolio construction technique can within real world constraints benefit a specific institutional investor who is subject to her own set of investment objectives and constraints.[17,18]

The framework of Molyboga, Baek, and Bilson provides the flexibility that is required for customization and accounts for the following real-life constraints:

1. Investment objectives vary substantially across institutional investors. Thus, the framework should allow for a broad range of performance metrics such as a Sharpe ratio, *certainty equivalent return*, t-statistics of alpha with respect to a benchmark or a factor model, *funding ratio* or probability of achieving a return in excess of a spending rate.

2. Institutional investors have their own investment constraints. Thus, the framework should be flexible enough to allow for incorporating customized investment constraints such as the minimum acceptable track record length, AUM, or the hedge fund style.

3. Investors generally target a discrete number of funds such as 5, 10, or 20 rather than hundreds typically considered in academic studies. The framework allows an investor to choose her own discrete number of funds that is kept fixed in the analysis.

4. Rebalancing frequencies vary across investors and hedge fund strategies as discussed in Section 2.2.2. Some of them may choose to rebalance portfolios monthly. That could be appropriate for a small number of highly liquid hedge fund strategies such as CTAs. Others may rebalance quarterly, semi-annually, or annually. The framework allows customization of the rebalancing frequency.

5. The empirical analysis accounts for all the biases discussed in detail in Section 2.2.1. For portfolios of CTAs, the framework imposes a one-month delay to mitigate the look-ahead bias. For example, if at the end of December, funds are selected for inclusion in a portfolio for January 2020, the funds' December returns are not available yet. Thus, the framework uses November 2019 returns in the analysis.

The framework follows the following steps:

- **First step: Data**. The dataset is chosen and adjusted for survivorship and backfill/incubation biases.
- **Second step: Eligible funds for each rebalance period**. For each rebalancing period (e.g., December 2019), the framework excludes all funds that fail to satisfy the investor's investment constraints such as the minimum track record length or AUM. This pool of "ALL" (SKILLED and NON-SKILLED) funds is used for the null hypothesis of no fund selection skill. Then for each fund selection technique considered, its own pool of funds is selected. For example, if the investor suspects that top quintile funds are "SKILLED" funds because they produce superior *ex-ante* performance, she will perform additional filtering to limit the "SKILLED" pool of funds to a sub-set of "ALL" funds that meets the fund selection criterion. Note that this analysis incorporates the one-month delay to ensure that the framework only relies on information that is available when the investment decision is made.
- **Third step: Single simulation**. The analysis starts with a selection of the discrete number of funds such as 5, 10, or 20 chosen by the investor from the "ALL" pool to form an "ALL" portfolio and the same number of funds from the "SKILLED" pool to form a "SKILLED" portfolio. The performance of both portfolios is recorded for the next rebalance period using a standard 1/N approach, highlighted in the 2009 study *Optimal versus Naive Diversification: How Inefficient is the 1/N Portfolio Strategy?*, or, alternatively, an equal-risk approach, discussed in detail in Section 5.2, and adjusted for the liquidation bias.[19] The funds remain in the portfolios as long as they continue satisfying the selection criteria by staying in their respective pools for the

following rebalance period. At the end of the simulation an "ALL" time-series and a "SKILLED" time-series that cover the complete out-of-sample period are recorded.

- **Fourth step: Large-scale simulation.** Since the methodology produces a large number of feasible portfolios in each period, it relies on a large-scale simulation approach designed to test hedge fund selection techniques in a way that is consistent with requirements of large institutional investors. That is accomplished by performing a large number of simulations (the number of simulations should be sufficiently large to have small sampling error and produce similar results each time the large-scale simulation analysis is repeated), such as 10,000, following the process described in the third step. The performance measure (e.g., Sharpe ratio) selected by the investor is used to calculate the performance of each "SKILLED" portfolio and each "ALL" portfolio. Because a large number of portfolios is considered, the framework produces a distribution of performance results (e.g., Sharpe ratio) for "SKILLED" portfolios and a distribution of performance results for the "ALL" portfolios that represent the null hypothesis. The evaluation of out-of-sample results is challenging primarily because simulation results are not independent as the returns of the same funds are used across many simulations; therefore, standard statistical tests are inappropriate. We will discuss how out-of-sample performance is evaluated further.

This framework is very flexible. It allows for a broad range of investment objectives and constraints, choice of the number of managers in a portfolio and can consider a large number of fund selection approaches.

In their 2017 paper *Assessing Hedge Fund Performance with Institutional Constraints: Evidence from CTA Funds*, Molyboga, Baek, and Bilson illustrated the framework by investigating performance persistence among CTAs, a subset of hedge funds that is primarily known for utilizing trend following or time-series momentum strategies in futures and options markets.[20] Institutional interest in CTAs has increased in response to the performance of these funds during the Global Financial Crisis with assets growing from US $131 billion in 2005 to almost US $390 billion in the second quarter of 2022, according to the BarclayHedge.[c]

The authors used the BarclayHedge database recommended in the 2021 paper *Hedge Fund Performance: Are Stylized Facts Sensitive to Which Database One Uses?* as the highest quality commercial CTA database.[21] The database included 4,909 active and defunct funds over the period December 1991–December 2013 with the out-of-sample period being January 1999–December 2013. Multi-advisors funds, funds with less than US $10 million in AUM and funds that failed to report net-of-fee returns were removed from the study. Moreover, the authors mitigated:

- Survivorship bias by including the graveyard database.
- Backfill bias using a combination of the AUM-based and truncation of a fixed number of returns approaches discussed in in Section 2.2.1.
- Liquidation bias by including a 1 percent penalty as suggested in the 1999 paper *The Performance of Hedge Funds: Risk, Return, and Incentives.*[22]

While the adjustments for survivorship and liquidation biases were adequate, the backfill bias adjustments were likely insufficient as shown in the 2014 paper *Fooling Some of the People All of the Time: The Inefficient Performance and Persistence of Commodity Trading Advisors.*[23] As discussed in Section 2.2.1, the optimal adjustment for the backfill bias includes relying on the first reported data field for returns starting in 2003 and adjustments from the 2017 study *Just a One Trick Pony? An Analysis of CTA Risk and Return* or the 2017 paper *The Fix is In: Properly Backing Out Backfill Bias* before 2003.[24,25]

In order to produce results that are relevant for institutional investors, they decided to investigate portfolios of 20 funds and incorporated two standard investment constraints by excluding funds who were in the bottom 30 percent of AUM or whose track record was less than 60 months old. The performance of the remaining funds was measured by calculating the t-statistic of alpha with respect to the BarclayHedge CTA index, a typical CTA benchmark, using data from the previous 60 months. The simulation framework used a lag of one

[c]https://www.barclayhedge.com/solutions/assets-under-management/cta-assets-under-management/

month to account for the delay in performance reporting of CTAs and employs 10,000 simulations.

A single simulation run resulted in several time-series that represented monthly out-of-sample returns of equally weighted (or equally risk-weighted) portfolios of randomly selected "ALL" CTAs and "SKILLED" CTAs chosen from the top quintile based on the t-statistic of alpha with respect to the CTA benchmark.

3.1.3 "ALL" Funds

The in-sample/out-of-sample framework followed a standard investment process of an institutional investor who makes allocation decisions at the end of each month. As discussed previously, the framework can handle any rebalancing frequency such as quarterly, semi-annual, and annual. The first decision was made in December 1998. Because of the delay of CTA reporting, the investor had information regarding fund returns and AUM through November 1998. As previously discussed, a delay of up to three months should be sufficient for most hedge fund strategies, but CTAs who tend to invest in highly liquid instruments tend to have shorter reporting delays. Therefore, the investor considered all funds that had a complete set of 60 months of returns between December 1993 and November 1998. First, the investor eliminated all funds in the bottom 30 percent of AUM among the funds considered. This flexible AUM threshold is more appropriate than a fixed AUM approach commonly used in the literature because the level of AUM increased substantially between 1999 and 2013.[26] Then the investor randomly chose 20 funds from the remaining pool of CTAs and allocated to them using two approaches. The first one is the equal nominal allocation (hereafter, EN), also known as the $1/N$ approach that allocates the same weight of $1/N$ to each asset i

$$w_i^{EN} = 1/N. \tag{3.1}$$

The second allocation approach is the equal volatility allocation (henceforth, EVA) which is an equal-risk approach that relies on volatility as a measure of risk. Sometimes this approach is also referred to as inverse volatility approach because it allocates to each asset i inversely to its volatility σ_i as follows:

$$w_i^{EVA} = \frac{1/\sigma_i}{\sum_{j=1}^{N} 1/\sigma_j}. \tag{3.2}$$

Volatility σ was estimated using sample standard deviations over the previous 60 months, allowing for a one-month reporting lag. The authors recommended using two approaches for robustness.

The return of both EN and EVA portfolios was calculated for January 1999 using the liquidation bias adjustment for the funds that liquidated during the month. At the end of January 1999, the pool of CTAs was updated and defunct constituents of the original portfolio were randomly replaced with funds from the new pool at which point the portfolio was rebalanced again using EN and EVA approaches. The process was repeated until the end of the out-of-sample period in December 2013. One single simulation resulted in two out-of-sample return stream between January 1999 and December 2013—one for the EN and the other one for the EVA approach.

3.1.4 "SKILLED" Funds

The in-sample/out-of-sample framework followed a very similar process when an institutional investor decides to limit the CTA pool only to those CTAs that rank in the top quintile based on the t-statistics of alpha with respect to the CTA benchmark. The first decision was made in December of 1998. Just as in the "SKILLED" fund selection case, the investor excluded all funds with less than 60 month track record and the bottom 30 percent of funds based on AUM. Then the investor ranked all funds using the t-statistic of alpha with respect to the CTA benchmark and only considered the funds that ranked in the top quintile. Appendix A.1 describes performance ranking of the funds. As previously mentioned, the investor can choose any fund selection approach whether based on AUM (e.g., select funds in the top quintile based on AUM), performance relative to factor models discussed further in this section, or any other selection criterion.

Once the "SKILLED" pool was determined, the investor randomly chose 20 funds from that pool and allocated to them using the EN and EVA approaches. The return of both EN and EVA portfolios was calculated for January 1999 using the liquidation bias adjustment for the funds that liquidated during the month. At the end of January 1999, the pool of CTAs was updated following the same procedure of ranking and the constituents of the original portfolio that failed to meet the selection criteria were randomly replaced with funds from the "SKILLED" pool at which point the portfolio was rebalanced again using EN and EVA approaches. The process was repeated until the end of the out-of-sample period in December 2013. A single simulation resulted in two out-of-sample return stream between January 1999 and December 2013—one for the EN and the other one for the EVA approach.

3.1.5 Evaluation of Out-of-sample Results

The simulations were run 10,000 times and the out-of-sample results included empirical distributions with 10,000 points (one per simulation). Table 3.1 presents the statistics of the Sharpe ratio distributions of the EN 20-fund portfolios selected from the "ALL" funds and the "SKILLED" funds. The improvement in the Sharpe ratios was meaningful with the mean Sharpe ratio going from 0.33 for all funds to 0.62 for the funds selected from the top quintile. Moreover, all five quartile statistics were also superior for the portfolios of the "SKILLED" funds. Since hedge fund investors may prefer stable portfolios, or incur turnover costs due to rebalancing frictions and additional due diligence, the "SKILLED" portfolios may need to be penalized within the framework to account for that.

Table 3.1 **Distributions of the Sharpe ratios.** The table reports the mean, standard deviation, and quartiles of the distribution of Sharpe ratios including the minimum, first quartile, median, third quartile, and maximum values of the EN 20-fund portfolios selected from "ALL" and "SKILLED" funds.

	ALL	SKILLED
Mean	0.33	0.62
StDev	0.07	0.06
Max	0.67	0.85
Third Quartile	0.38	0.66
Median	0.33	0.62
First Quartile	0.28	0.58
Min	0.05	0.37

However, the evaluation of out-of-sample results is challenging because simulation results are not independent as the same funds are used across many simulations. Therefore, standard statistical tests are inappropriate and the framework has to rely on bootstrapping tests, introduced in the 1979 study *Bootstrap Methods: Another Look at the Jackknife* by Bradley Efron and the 1983 study *A Leisurely Look at the Bootstrap, the Jackknife, and Cross-validation* by Bradley Efron and Gail Gong.[27,28] Molyboga, Baek, and Bilson provided a detailed description of how bootstrapping can be used to compare means and test for *stochastic dominance*.[29] Stochastic dominance is a comprehensive measure of performance that considers the entire distribution of returns rather just mean and variance used in standard mean-variance analysis. Second-order stochastic dominance is particularly attractive because if portfolio A dominates portfolio B, that implies that all risk-averse investors, regardless of their utility functions, should unanimously prefer A to B.[30]

The aforementioned 2017 paper *Assessing Hedge Fund Performance with Institutional Constraints: Evidence from CTA Funds* found strong persistence in the performance of top-performing CTA funds using second order stochastic dominance tests.[31] While their finding is consistent with those of the authors of the 2021 paper *Hedge Fund Performance: Are Stylized Facts Sensitive to Which Database One Uses?* who reported performance persistence among hedge funds, their conclusion is potentially driven by backfill bias since they didn't account for it using the first reported date.[32]

Once the dataset is correctly adjusted for the backfill bias, the large scale framework of Molyboga, Baek, and Bilson provides a robust and flexible methodology for evaluation of fund selection approaches with real-life constraints. The framework is customizable to the specific investment objectives and constraints of investors.

3.2 FACTORS AND FACTOR SELECTION

It is widely accepted in academic literature to evaluate performance of investments relative to systematic factors because the finance theory suggests that only systematic sources of risk are compensated with higher expected returns. The systematic sources of risk can be either macroeconomic, as proposed in the 1986 study *Economic Forces and the Stock Market* by Nai-Fu Chen, Richard Roll, and Stephen Ross, or investable portfolios, as illustrated in the 1996 study *Multifactor Explanations of Asset Pricing Anomalies* by Eugene Fama and Kenneth French.[33,34]

The Fung-Hsieh seven-factor model has been the primary benchmark model for evaluating hedge fund performance since it was proposed in the 2004 paper *Hedge Fund Benchmarks: A Risk-based Approach* by William Fung and David Hsieh.[35,36] The seven-factor model includes two equity-oriented risk factors, two bond-oriented risk factors, and three trend following factors:

1. Equity market factor. Monthly returns of the S&P 500 total return index.
2. Size spread factor. Monthly returns of the Russell 2000 total return index minus monthly returns of the S&P 500 total return index. Earlier studies, such as the 2001 paper *The Risk in Hedge Fund Strategies: Theory and Evidence from Trend Followers*, used the Wilshire Small Cap 1750 minus the Wilshire Large Cap 750 monthly returns.[37]
3. Bond market factor. The monthly change in the 10-year Treasury constant maturity yield.
4. Credit spread factor. The monthly change in the Moody's Baa yield minus 10-year Treasury constant maturity yield.
5. Bond trend following factor.
6. Currency trend following factor.
7. Commodity trend following factor.

The eight-factor Fung-Hsieh model includes an additional factor—an emerging market factor—that is, often proxied by the MSCI Emerging Market total return index. However, the aforementioned 2001 paper *The Risk in Hedge Fund Strategies: Theory and Evidence from Trend Followers* used the IFC Emerging Market total return index.[38]

In their original 2004 study, Fung and Hsieh showed that the seven-factor model explained up to 80 percent of monthly return variations. That finding established the seven-factor model as the primary benchmark model for evaluating hedge fund performance. For example, it has been used to examine skill and performance persistence in the aforementioned 2004 paper *Analysis of Hedge Fund Performance*, the 2007 study *Do Hedge Funds Deliver Alpha? A Bayesian and Bootstrap Analysis*, the 2010 paper *Do Hot Hands Exist Among Hedge Fund Managers? An Empirical Evaluation*, and the 2021 paper *Hedge Fund Performance: Are Stylized Facts Sensitive to Which Database One Uses?*[39,40,41,42] It was also heavily used in other hedge fund research areas, such as evaluating managerial incentives in the 2009 paper *Role of Managerial Incentives and Discretion in Hedge Fund Performance*.[43]

3.2.1 Issues with Standard 7-Factor Fung-Hsieh Model

However, recent research highlighted issues with the standard Fung-Hsieh model. For example, Nicolas Bollen in his 2013 study *Zero-R^2 Hedge Funds and Market Neutrality* showed that the Fung-Hsieh model suffered from an omitted factor issue.[44] He showed that roughly a third of all hedge funds had an R^2 that was close to zero and, thus, their risk was purely idiosyncratic. Since idiosyncratic risk can be diversified away in large portfolios, portfolios of zero-R^2 funds should be associated with low risk. However, Bollen found that zero-R^2 portfolios had abnormally high volatility and probability of failure suggesting that the Fung-Hsieh model failed to fully capture systematic risk.

Moreover, the aforementioned 2014 paper *Fooling Some of the People All of the Time: The Inefficient Performance and Persistence of Commodity Trading Advisors* by Geetesh Bhardwaj, Gary Gorton, and Geert Rouwenhorst criticized the option-based trend factors in the Fung-Hsieh model as inefficient replicators of trend that produce an upward bias in alphas.[45]

This criticism can be easily validated by considering the cumulative performance of the Fung-Hsieh trend factors available at David Hsieh's Hedge Fund Data Library.[46] Figure 3.1 displays the cumulative performance of five primitive trend factors PTFSBD (bonds), PTFSFX (foreign exchange), PTFSCOM (commodities), PTFSIR (interest rates), and PTFSSTK (stocks).

The negative performance of the Fung-Hsieh factors is in sharp contrast with the superior performance of the time-series momentum strategy, introduced in the 2012 paper *Time-Series Momentum* by Tobias Moskowitz, Yao Ooi, and Lasse Pedersen.[47] Time-series momentum takes a long exposure in a security, if its cumulative lagged 12-month return is positive, and a short exposure, if the return is negative, holds positions for a month, and then rebalances the portfolio based on the most recent cumulative 12-month return. Table 3.2 compares the performance of the trend strategies. While the Sharpe ratios of the Fung-Hsieh factors ranged between −0.90 and −0.023, the time-series momentum strategy delivered a high Sharpe ratio of 0.86 over the same period between 1994 and 2020.

The aforementioned 2021 paper *Hedge Fund Performance: Are Stylized Facts Sensitive to Which Database One Uses?* by Juha Joenvaara, Mikko Kauppila, Robert Kosowski, and Pekka Tolonen

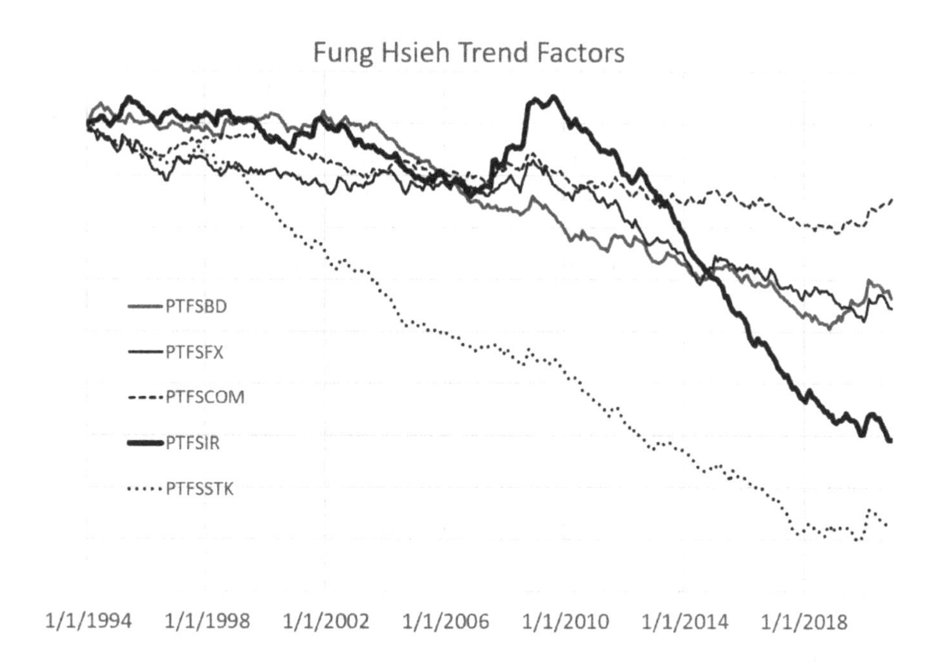

Figure 3.1 Cumulative performance of five Fung-Hsieh primitive trend factors: January 1994–December 2020. The figure displays performance of the five factors: PTFSBD, PTFSFX, PTFSCOM, PTFSIR, and PTFSSTK. The scale is logarithmic.

Table 3.2 **Performance of five primitive trend strategies of Fung-Hsieh and time-series momentum: January 1994–December 2020.** The table reports the annualized excess returns, annualized standard deviations, and the Sharpe ratios of PTFSBD, PTFSFX, PTFSCOM, PTFSIR, PTFSSTK, and TSMOM.

	PTFSBD	PTFSFX	PTFSCOM	PTFSIR	PTFSSTK	TSMOM
Ann Ret	−24.80%	−26.09%	−11.61%	−40.43%	−48.16%	10.84%
Ann StDev	57.97%	69.04%	50.38%	85.48%	53.75%	12.56%
Sharpe	−0.43	−0.38	−0.23	−0.47	−0.90	0.86

introduced an alternative benchmark model for evaluating hedge fund performance.[48] Their factor model includes:

- The global Carhart four-factor model originally introduced in the 1997 study *On Persistence in Mutual Fund Performance* by Mark Carhart.[49] The model includes the market, size and value factors defined for global portfolios in the 2012 study *Size, Value, and Momentum in International Stock Returns* by Eugene Fama and Kenneth French and the global momentum portfolio from the 2013 paper *Value and Momentum Everywhere* by Clifford Asness, Tobias Moskowitz, and Lasse Pedersen.[50,51]

- The time-series momentum factor from the 2012 paper *Time-Series Momentum* by Tobias Moskowitz, Yao Ooi, and Lasse Pedersen.[52]

- The Pastor-Stambaugh liquidity factor introduced in the 2003 paper *Liquidity Risk and Expected Stock Return* by Lubos Pastor and Robert Stambaugh.[53]

- The betting-against-beta factor from the 2014 study *Betting Against Beta* by Andrea Frazzini and Lasse Pedersen. [54]

It is worth noting that factor returns are typically provided gross of trading costs. This is particularly important for high-turnover factors such as the Pastor-Stambaugh liquidity factor. Since hedge fund returns are net of trading costs, factor regressions tend to understate hedge fund alphas.

Due to absence of an established robust model for benchmarking hedge funds, we propose expanding the list of potential factors and then applying a factor selection approach. Many factor selection approaches in finance emerged from machine learning literature. For example, the 2020 study *Empirical Asset Pricing via Machine Learning* by Shihao Gu, Bryan Kelly, and Dacheng Xiu recommended two standard machine learning techniques: the *least absolute shrinkage and selection operator*, or LASSO, introduced in the 1996 paper *Regression Shrinkage and Selection via the Lasso* by Robert Tibshirani, and the *Elastic Net* approach, introduced in the 2005 paper *Regularization and Variable Selection via the Elastic Net* by Hui Zou and Trevor Hastie.[55,56,57] Section 3.2.5 provides a comprehensive discussion of factor selection.

In the next three sections, we discuss three factors—volatility premium, short-term momentum, and term premium—that extend standard factor models along three dimensions: volatility, time-frame, and forward curve. Researchers may want to consider adding these factors when evaluating different types of hedge funds.

3.2.2 Volatility Premium

As previously discussed, the 2013 study *Zero-R^2 Hedge Funds and Market Neutrality* showed that the Fung-Hsieh model suffered from an omitted factor issue.[58] Since hedge fund strategies tend to have a profile that is similar to short options, it seems reasonable to consider volatility selling strategies.

We use the CSI data to construct a simple volatility selling strategy that sells VIX futures and periodically rolls the position to the most active contract. Table 3.3 presents the summary statistics of the volatility selling strategy. While its volatility is high, it has produced an attractive Sharpe ratio of 0.91 over the period March 2004–December 2020.

Table 3.3 **Performance of a volatility selling strategy: March 2004–December 2020.** The table reports the annualized return and standard deviation of excess returns of the volatility selling strategy that shorts VIX futures, and its Sharpe ratio.

	Volatility Selling
Annualized Return	30.59%
Annualized Standard Deviation	33.59%
Sharpe Ratio	0.91

Figure 3.2 displays the cumulative performance of a volatility selling strategy that shorts VIX futures.

The strategy performed well over this period. However, it experienced large drawdowns during market downturns. For example, the drawdown was almost 70 percent during the Global Financial Crisis in the last quarter of 2008 and it reached about 55 percent in the first quarter of 2020 in response to the pandemic.

3.2.3 Short-term Momentum

Another example of a factor that may be complementary to the standard factor models is short-term momentum introduced in the 2020 study *Short-Term Trend: A Jewel Hidden in Daily Returns* by

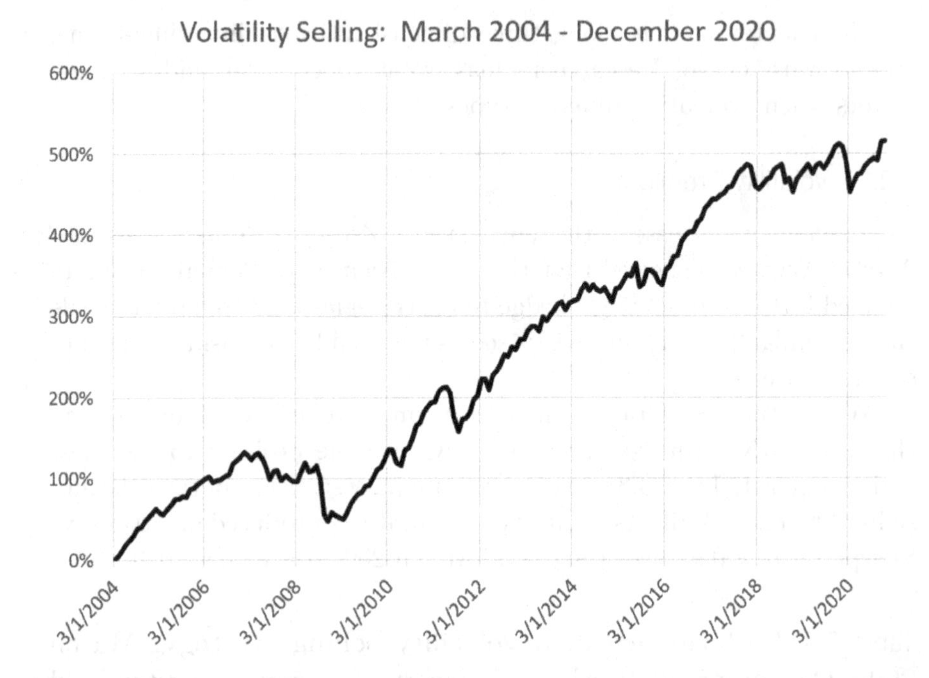

Figure 3.2 Cumulative performance of a volatility selling strategy: March 2004–December 2020. The figure displays performance of a volatility selling strategy that shorts VIX futures.

Marat Molyboga, Larry Swedroe, and Junkai Qian.[59] The short-term momentum factor is an extension of the standard time-series momentum described earlier. Time-series momentum, introduced in the 2012 paper *Time-Series Momentum,* takes a long exposure in a security, if its cumulative lagged 12-month return is positive, and a short exposure, if the return is negative, holds positions for a month, and then rebalances the portfolio based on the most recent cumulative 12-month return.[60] This particular strategy is often denoted as 12-1 momentum. The 2017 study *A Century of Evidence on Trend-Following Investing* by Brian Hurst, Yao Ooi, and Lasse Pedersen showed that 3-1 and 1-1 time-series momentum strategies that rely on 3-month and 1-month lagged returns with monthly rebalancing, respectively, positively contributed to the original strategy within an equally-weighted portfolio.[61]

Molyboga, Swedroe, and Qian examined whether using daily returns to generate signals and rebalancing intra-month could further improve the aggregate performance of time-series momentum strategies and whether that improvement could be captured by investors after transaction costs. Since there are approximately 21 trading days in a month and 252 trading days in a 12-month period, the 12-1 monthly momentum strategy was approximated with the 252-21 daily momentum strategy. The 252-21 strategy uses a cumulative lagged return over 252-days to generate a signal, holds the position for 21 days, at which point it generates the next signal. The authors also examined the 63-21 and 21-21 daily time-series momentum strategies that resemble the 3-1 and 1-1 monthly time-series momentum strategies from the just cited 2017 study *A Century of Evidence on Trend-following Investing.* Molyboga, Swedroe, and Qian found that the daily strategy performed similarly to the monthly strategies.

Moreover, the authors introduced the 21-5 daily short-term momentum strategy. The five-day rebalancing frequency was chosen to match a weekly rebalancing frequency. Unlike the other three strategies, the short-term momentum strategy cannot be approximated using monthly returns.

Molyboga, Swedroe, and Qian found that the short-term momentum strategy had an attractive Sharpe ratio of 0.83 and moderate correlations to the other three strategies ranging between 0.21 with the longer-term 252-21 strategy and 0.61 with the shorter-term 21-21 strategy. Moreover, they discovered a positive marginal contribution of short-term momentum on the performance of time-series momentum portfolios as standalone investments and as diversifiers to stock portfolios. Since short-term momentum is a high-*turnover* strategy, the authors investigated the robustness of the performance improvement to transaction costs. They found that the degree of improvement was highly dependent on the quality of execution, and, therefore, prudent hedge fund managers needed to invest in their execution infrastructure and algorithms to benefit from this attractive high-turnover strategy.

Therefore, since most factor models are constructed using monthly returns and hedge funds often generate signals using daily returns, incorporating strategies based on daily returns may be better suited for evaluation of some hedge funds.

3.2.4 Term Premium in Commodities

Although using volatility and short-term momentum factors can help in explaining the performance of some hedge fund strategies, there is another gap in hedge fund benchmarks. Popular hedge fund benchmarks such as the Fung-Hsieh seven-factor model include a term premium but it is only limited to fixed income. Hedge funds may want to capture the term premium that exists in other asset classes. For example, the 2014 study *An Anatomy of Commodity Futures Risk Premia* reported a persistent and statistically significant term premium return in commodities.[62]

The 2018 study *Benchmarking Commodity Investments* by Jesse Blocher, Ricky Cooper, and Marat Molyboga introduced a simple implementable calendar spread strategy that went long further out contracts and shorted nearby contracts.[63] Blocher, Cooper, and Molyboga showed that the term premium in commodities had been significant since the financialization of commodities around 2003–2004, when institutional investors started recognizing commodities as a distinct asset class that should be included in global investment portfolios. Such term premium strategies may help evaluate hedge funds that actively trade forward curves across asset classes outside of fixed income.

3.2.5 Factor Selection

Empirical studies often rely on two types of regressions. The first one is the *ordinary least squares* (OLS) regression that assumes that residuals are not autocorrelated (zero serial correlation) and have constant volatility (homoskedasticity). The second approach is the regression with the *Newey-West adjustment*, introduced in the 1987 paper *A Simple, Positive Semi-definite, Heteroskedasticity and Autocorrelation Consistent Covariance Matrix* by Whitney Newey and Kenneth West that relaxes the OLS assumptions and allows for non-zero autocorrelation and non-constant volatility (*heteroskedasticity*) of residuals.[64] Although both approaches result in identical estimates of the intercepts and slope coefficients, they produce different t-statistics. Since hedge fund returns tend to exhibit positive serial correlation and heteroskedasticity, OLS estimates tend to overstate the statistical significance of regression coefficients.

Although Newey-West adjustment accounts for heteroskedasticity and serial correlation, it works poorly when the number of explanatory variables is large, which is an important obstacle because evaluation of hedge funds often requires considering a large number of potential factors. As discussed in the 2019 paper *Artificial Intelligence in Finance*, quickly growing machine learning is reshaping the financial services industry since conventional econometric approaches are designed to rely on relatively small number of factors whereas machine learning methods are designed for utilizing a large number of factors for predictive accuracy.[65]

The aforementioned 2020 study *Empirical Asset Pricing via Machine Learning* by Shihao Gu, Bryan Kelly, and Dacheng Xiu pointed out that standard econometric approaches produce forecasts that are highly unstable out-of-sample when the number of potential predictors is large relative to the length of financial time series and recommended two standard machine learning techniques: the LASSO and the Elastic Net approaches.[66]

The LASSO approach attempts to select only a sub-set of factors that are relevant by penalizing all non-zero regression coefficients and eliminating spurious relationships. LASSO was used to investigate lead-lag relationships among international markets and industries in the 2013 study *International Stock Return Predictability: What is the Role of the United States?* and the 2019 study *Industry Return Predictability: A Machine Learning Approach.*[67,68] It was also used for characteristic-based factor selection in three 2020 studies *A Transaction-cost Perspective on the Multitude of Firm Characteristics, Taming the Zoo: A Test of New Factors,* and *Dissecting Characteristics Nonparametrically.*[69,70,71]

In their 2019 paper *Sentiment Indices and Their Forecasting Ability* and 2021 paper *Market Timing Using Combined Forecasts and Machine Learning,* David Mascio and Frank Fabozzi showed that the LASSO approach was more effective at selecting sentiment factors that collectively predict stock market returns than the conventional sentiment index and kitchen sink logistic regression models.[72,73]

The 2005 paper *Regularization and Variable Selection via the Elastic Net* by Hui Zou and Trevor Hastie showed that when potential predictors are correlated, the Elastic Net approach that penalizes non-zero regression coefficients differently results in superior prediction accuracy.[74] The aforementioned 2020 study *Empirical Asset Pricing via Machine Learning* explained that LASSO is effective at factor selection but Elastic Net also mitigates the issue of estimated coefficients being too large.[75]

The 2012 paper *Sparse Models and Methods for Optimal Instruments with an Application to Eminent Domain* introduced the Post-LASSO estimator that relies on LASSO for factor selection and then uses those factors to re-estimate the coefficients with an OLS methodology.[76] The aforementioned 2020 study *Taming the Zoo: A Test of New Factors* suggested that the choice of method in a given context depends on the underlying model assumption.[77] Thus, both the LASSO and Elastic Net approaches can be considered for factor selection. Appendix A.2 provides a technical description of OLS, LASSO, and Elastic Nets.

3.2.6 Quantitative and Qualitative Factors

In October 2020, the Chartered Alternative Investment Analyst (CAIA) Association presented the results of the study *Alternative Investment Due Diligence: A Survey on Key Drivers for Manager Selection* by Mark Rzepczynski and Keith Black.[78] It included a comprehensive survey of 233 investors and 111 managers with the goal of identifying the factors that are important for alternative investment manager selection. The investor respondents had extensive experience of manager selection and due diligence with about 45 percent having more than 10 years of experience and only 10 percent having less than two years of experience. The due diligence process was quite complex with two to four direct due diligence meetings held over a three- to nine-month period, on average.

The authors reported that:

- Alternative investment manager selection is a complex process that relies on both quantitative and qualitative analysis that cannot be captured through specific empirical measures of skill.
- Manager skill assessment for alternative investment is considered more difficult than selecting traditional investment managers and requires greater analysis of the philosophy, culture, and processes of the manager.
- Qualitative factors for alternative manager skill assessment are as important or more important than the quantitative assessment of alternative manager.
- Operational due diligence can dominate or override the assessment of investment skill and is critical to the manager selection process.
- The manager selection process is tailored to the strategy being reviewed. Thus, the specific issues or factors involved with choosing a private equity manager are very different from factors associated with a systematic hedge fund manager, for example.

3.2.7 Operational Due Diligence of Digital Asset Funds

As discussed in the 2022 study *Operational Due Diligence on Cryptocurrency and Digital Asset Funds,* the cryptocurrency space is transitioning from retail investors who were the early adopters to institutional investors who often allocate via third-party external fund managers, such as hedge fund managers, that invest in digital assets.[79] The authors discussed two main reasons why institutional investors were concerned about the operational risk of crypto hedge fund managers:

- Crypto hacks and frauds. For example, hackers stole around $400 million from Mt. Gox, a Tokyo-based bitcoin exchange, in 2014, and around $500 million from Coincheck, a Japanese firm, in 2018.[80]

- A perception that bitcoin is primarily used by criminals. For example, in her testimony to the Senate Finance Committee on January 19, 2021, Janet Yellen, the Secretary of the Treasury, said: "I think many cryptocurrencies are used, at least in a transaction sense, mainly for illicit financing. And I think that we really need to examine ways in which we can curtail their use, and make sure that anti-money laundering doesn't occur through those channels." Although cryptocurrencies are often used for ransomware payments, recent studies report that the criminal share of all cryptocurrency activity represented only 0.34 percent of transaction volume in 2020.[81]

Thus, as hedge funds expand to the crypto space, operational due diligence process should be revised accordingly. The authors discussed three important operational due diligence trends exhibited by institutional investors:

- Emergence of crypto-specific specialization of operational due diligence relative to other types of alternative investments. This trend is driven by the rapid innovation in the digital asset space that leads to new types of coins, token types (such as non-fungible tokens), and DeFi projects.

- Scrutiny of crypto custody arrangements. The early stage involved self-custody with cold wallets on personal computers or external hard drives, which was vulnerable to the risks of hacking, hardware failure, and password recovery. Self-custody has been gradually replaced with hybrid solutions, such as hardware wallets with two-factor authentication, and the third-party custodian, a standard solution in alternative investments. The third-party solution was employed by 52 percent of crypto hedge funds in 2019. Thus, the quality of third-party custodians has to be carefully evaluated during operational due diligence.

- Integration of operational due diligence and background investigation. Operational due diligence of crypto hedge funds several categories of background analysis, such as criminal, regulatory, and litigation checks. The background analysis is particularly helpful with addressing the illicit financing concern highlighted by Yellen.

3.3 KEY TAKEAWAYS

Following are the key takeaways from this chapter:

- Hedge fund selection should be customized to the specific objectives and constraints of investors, and identify hedge fund portfolios with a feasible number of funds with a positive aggregate marginal impact on the existing portfolios.

- The established Fung-Hsieh factor model has serious flaws. A factor model used for hedge fund evaluation should be based on the risk premia captured by the hedge fund. Volatility premium, short-term momentum, and term premium are a few potential candidates that are not actively used today. Since factors are usually reported gross of trading costs whereas hedge fund performance is net of trading costs, factor regressions tend to understate hedge fund alphas.

- When the number of potential factors is large, regularized techniques such as LASSO and Post-LASSO solve the problem of factor selection in hedge fund performance evaluation.

- Hedge fund manager selection considers quantitative and qualitative factors. Digital asset funds must go through specialized operational due diligence.

CHAPTER 4

Performance Persistence

"I think we consider too much the good luck of the early bird and not enough the bad luck of the early worm."—Franklin D. Roosevelt.[a]

"Past performance does not necessarily predict future results."[b]

After discussing issues related to evaluating hedge funds solely based on past performance, we turn to considering strategies that we believe can improve the likelihood of better future outcomes. This chapter provides a comprehensive overview of critical topics of predictive manager selection that include separation of luck from skill and performance persistence. It also includes a framework for combining quantitative and qualitative factors within a Bayesian framework.

4.1 PERFORMANCE PERSISTENCE

The topic of performance persistence is challenging, and we attempt to discuss two important questions:

- Does skill exist or does luck explain the cross-sectional variation in performance?
- Can we predict which hedge funds will outperform?

[a]Franklin D. Roosevelt (FDR) was the 32nd U.S. president.
[b]SEC, "Investor Bulletin: Performance Claims", Sept. 15, 2022.

DOI: 10.1201/9781003175209-4

4.1.1 Does Skill Exist?

It is unclear whether skill exists because luck can play a large role in outcomes. There are several issues associated with that. In statistics, inferences are made by comparing the value of a statistic to its distribution under the null hypothesis. If someone gives you a coin and asks you to determine whether it is a regular coin, you can flip it 24 times and use a binomial distribution table to determine the likelihood of seeing the outcome given the null hypothesis of the 50-50 odds. For example, if you observe 19 heads, the table will show you that the probability of seeing at least 19 heads is less than 1 percent and you should reject the null hypothesis of the 50-50 odds. By contrast, if you observe 15 heads, the probability would be close to 15 percent and you should not reject the null hypothesis. Rather than using a binomial table, a researcher can use a computer program with a random number generator to simulate the distribution. This simulation-based approach is called Monte Carlo simulation or bootstrapping.

The problem arises when there are many simultaneous experiments run instead of a single one. For example, if we perform the above experiment 10,000 times with a standard 50-50 coin, we expect to see 33 series with at least 19 heads. This issue is called *multiple hypothesis testing* which can lead to false discoveries. Similarly, if the t-statistic of alpha of 2 is sufficiently high when a single hedge fund is evaluated, observing a cross-section of 10,000 hedge funds will likely lead to seeing a large number of hedge funds with the t-statistic of alpha that exceeds 2 purely due to chance even if all hedge funds have a zero true alpha.

There are several techniques that are designed to identify evidence of skill. We consider two approaches here:

- A bootstrap approach as recommended in the 2006 study *Can Mutual Fund "Stars" Really Pick Stocks? New Evidence from a Bootstrap Analysis*, the 2007 study *Do Hedge Funds Deliver Alpha? A Bayesian and Bootstrap Analysis*, and the 2010 paper *Luck versus Skill in the Cross-Section of Mutual Fund Returns.*[1,2,3]

- A *false discovery rates* approach as demonstrated in the 2010 paper *False Discoveries in Mutual Fund Performance: Measuring Luck in Estimated Alphas.*[4]

4.1.1.1 A Bootstrap Approach

The aforementioned 2007 study *Do Hedge Funds Deliver Alpha? A Bayesian and Bootstrap Analysis* by Robert Kosowski, Narayan Naik, and Melvyn Teo and the 2010 paper *Luck versus Skill in the Cross-Section of Mutual Fund Returns* by Eugene Fama and Kenneth French examined the existence of skill among mutual funds and hedge funds, respectively.[5] As discussed by Fama and French, the goal of bootstrapping is to determine whether the cross-section of alpha estimates suggests that true alpha is zero for all funds or whether there is evidence of nonzero true alpha.

Hedge fund investors are especially interested in the tails of the cross-section of alpha estimates because they want to invest in hedge funds with true positive alpha and avoid hedge funds with true negative alpha. For example, if we observe that 100 out of 10,000 funds have the t-statistic of alpha that exceeds 2, does that mean that some of those funds have positive alpha? In order to answer this question, it is essential to know how many funds out of 10,000 zero alpha funds are expected to have the t-statistic of alpha greater than 2. If the theoretical number is equal to 40, then the dataset has 60 extra funds with a high value of the t-statistic of alpha, which can be interpreted as evidence of skill. However, if the theoretical number is close to 100, that indicates that the high values of the t-statistic of alpha can be explained by chance.

A bootstrapping procedure creates a simulated distribution of the t-statistic of alpha that can be compared to the actual distribution to draw conclusions about existence of skill. Appendix B.1 provides a detailed description of bootstrapping implementation steps. Using the bootstrap approach, Kosowski, Naik, and Teo found strong evidence of positive skill relative to the Fung-Hsieh model among hedge fund managers.[6]

4.1.1.2 A False Discovery Rates Approach

As discussed in the 2010 paper *False Discoveries in Mutual Fund Performance: Measuring Luck in Estimated Alphas* by Laurent Barras, Olivier Scaillet, and Russ Wermers, bootstrap approaches are used to test the hypothesis that all fund alphas are equal to zero and, therefore, they can determine whether positive and negative alpha funds exist. However, they fail to answer two important questions:

1. How many fund managers have positive or negative alpha?
2. What is the estimate of that positive or negative alpha (or the t-statistic of alpha)?[7]

Since the true alphas of funds are not observable, researchers investigate the cross-section of estimated alphas. However, simply counting the number of large estimated alphas fails to adequately account for luck because many funds may have high alphas by luck alone. For example, if all funds have true zero-alpha, 5 percent of 1,000 funds (50 funds) are expected to have positive estimated alphas that are significant at the 5 percent level. Barras, Scaillet, and Wermers call them "false discoveries"—"funds with significant estimated alphas, but zero true alphas." While the bootstrap approach compares the actual and simulated distributions of the t-statistics of alpha, the false discovery approach focuses on the distribution of their *p-values* instead. If all funds have true alpha of zero, their p-values should be uniformly distributed between 0 and 1.[8] Therefore, a *histogram* of estimated p-values can be compared to a uniform distribution and a disproportionately high share of low p-values serves as evidence of either positive alpha or negative alpha funds. Then the methodology considers the distribution of the estimated t-statistics of alpha to calculate the proportion of positive alpha funds and the proportion of negative alpha funds. The method assumes that all positive alpha funds have the same positive true alpha and all negative alpha funds have the same negative true alpha.

The authors applied their false discovery approach to 2,076 U.S. open-ended mutual funds for the period between 1975 and 2006. They found that 75.4 percent of funds were zero-alpha funds, 24.0 percent of funds had negative alpha, and only 0.6 percent had positive alpha. The false discovery rate approach can also be applied to hedge funds. Appendix B.2 includes detailed implementation steps of the false discovery approach.

4.1.2 Performance Evaluation with a Noise Reduced Alpha Approach

In the 2018 study *Detecting Repeatable Performance,* Campbell Harvey and Yan Liu attempted to pool information from the cross-sectional distribution of alphas to improve forecasts of individual fund alphas.[9]

The 2007 study *Do Hedge Funds Deliver Alpha? A Bayesian and Bootstrap Analysis* considered the world with all hedge funds with a true zero alpha.[10] The 2010 paper *False Discoveries in Mutual Fund*

Performance: Measuring Luck in Estimated Alphas assumed that the hedge fund managers had either zero-alpha, negative alpha, or positive alpha funds with each fund within a group sharing the same true alpha.[11]

Harvey and Liu argued that limiting the potential set of hedge fund alphas to only three values was too restrictive and introduced a methodology with a much broader set of potential alpha values. The authors examined their novel methodology using a simulation study that matched many essential features of mutual fund data. They found that their approach had higher forecasting accuracy of alphas than all alternative methods considered in the study. The authors also found that the proportion of mutual funds with true positive alpha was closer to 10 percent than to the previously reported 0–1 percent. Harvey and Liu stated: "The very low proportion found in previous research is due to the high level of estimation uncertainty associated with a fund-by-fund analysis. Our framework provides a more powerful procedure to identify funds with small positive alphas by directly modeling the underlying alpha population." Appendix B.3 includes a technical discussion and implementation steps of the false noise reduced alpha approach.

4.1.3 Performance Evaluation with Seemingly Unrelated Assets

Hedge fund performance evaluation is very challenging because of having to rely on short track records to draw conclusions regarding skill. If returns of a hedge fund with a short track record are regressed on the hedge fund factors, a standard OLS (Ordinary Least Squares) estimate of alpha can be noisy. The 2002 paper *Mutual Fund Performance and Seemingly Unrelated Assets* by Lubos Pastor and Robert Stambaugh introduced an elegant Bayesian Pastor-Stambaugh approach with seemingly unrelated assets designed to improve the estimation quality of alphas and applied it to mutual funds.[12]

They considered mutual fund benchmarks as "seemingly unrelated assets." Pastor and Stambaugh showed that they could draw additional information in seemingly unrelated assets to mitigate the short sample problem and improve the accuracy of performance estimates. They examined the performance of mutual funds relative to their benchmarks and factors during their short track records and the performance of mutual fund benchmarks relative to the factors over longer time period since benchmarks had longer track records than individual funds.

The aforementioned 2007 paper *Do Hedge Funds Deliver Alpha? A Bayesian and Bootstrap Analysis* by Kosowski, Naik, and Teo applied the Bayesian Pastor-Stambaugh approach to hedge funds.[13] The authors combined monthly hedge fund returns from the CSFB/Tremont TASS, Hedge Fund Research, Center for International Securities and Derivatives Markets, and Morgan Stanley Capital International data sets for the periods from 1990 to 2002. They considered hedge fund benchmarks as "seemingly unrelated assets" and found that the Pastor-Stambaugh approach drastically improved the predictability in hedge fund returns relative to the standard OLS approach. When Kosowski, Naik, and Teo sorted hedge funds based on their two-year past Bayesian alphas, they found that the top decile hedge fund portfolio outperformed the bottom decile hedge fund portfolio by approximately 5.81 percent per annum, which is significant in economic and statistical terms with the t-statistic of 2.65. By contrast, the standard OLS approach produced a 0.73 percent annual spread between the top and bottom decile hedge fund portfolios with the corresponding t-statistic of 0.24.

Kosowski, Naik, and Teo showed that their performance persistence findings were not driven by alternative explanations such as serial correlation in fund returns and concluded that top hedge fund managers possessed asset selection skill. Appendix B.4 includes technical details and implementation steps of the seemingly unrelated assets approach.

4.1.4 Performance Evaluation with Decreasing Returns to Scale

While most performance evaluation approaches solely rely on historical returns, the theoretical model of Berk and Green, and the empirical investigation in the 2012 study *The Life Cycle of Hedge Funds: Fund Flows, Size, Competition, and Performance* by Mila Getmansky point to the decreasing returns to scale assumption—fund alphas diminish with asset growth.[14,15]

The 2021 paper *Marketing Mutual Funds* by Nikolai Roussanov, Hongxun Ruan, and Yanhao Wei introduced a novel approach to performance evaluation with decreasing returns to scale.[16] Their model assumed that a fund's alpha was determined by the skill of the fund manager and the fund's asset size. Unlike Berk and Green who assumed that managerial skill didn't change with time, Roussanov, Ruan, and Wei assumed that the skill of a fund manager slowly reverted to the industry-average skill level. In their model, the asset growth had a negative impact on alpha because of the decreasing returns to scale. They imposed a Bayesian framework to estimate model parameters such as the industry-average skill level and the skill reversion speeds as well as the fund alphas.

Roussanov, Ruan, and Wei applied the model to a sample of 2,285 well-diversified actively managed domestic equity mutual funds from the U.S. covering the period 1964–2015 from CRSP and Morningstar, and concluded that marketing of mutual funds was almost as important for attracting assets as performance and fees. They found that a 1 *basis point* increase in marketing expenses led to a 1 percent increase in a fund's AUM.

Roussanov, Ruan, and Wei discovered that marketing produced asset misallocation. When they sorted mutual funds on their net-of-fee managerial skill, they found that the top decile funds were too small—their average AUM of $936 million was significantly smaller than the model implied AUM of $7.3 billion required to reduce net-of-fee alphas to zero. By contrast, the bottom eight deciles were too large—their excessive AUM was leading to negative net-of-fee alphas.

The 2022 paper *Mutual Fund Flows and Performance in (Imperfectly) Rational Markets?* by Nikolai Roussanov, Hongxun Ruan, and Yanhao Wei used a similar Bayesian approach to investigate whether mutual fund investors chase performance or rationally update their beliefs.[17] Roussanov, Ruan, and Wei applied the model to a sample of 2,377 well-diversified actively managed domestic equity mutual funds from the U.S. covering the period 1965–2014 from CRSP and Morningstar and found that retail investors and to a lesser degree institutional investors exhibited behavioral biases. The investors were overly optimistic about manager skill and chased performance.

4.1.5 Identifying Hedge Fund Skill with Peer Cohorts

The 2021 study *Identifying Hedge Fund Skill by Using Peer Cohorts* by David Forsberg, David Gallagher, and Geoffrey Warren introduced another approach to detecting skill based on peer cohorts.[18] The method uses correlations to form peer groups of hedge funds and then selects managers based on their performance relative to their peer groups. The approach is not subject to the omitted variables problem of hedge fund factor models highlighted in the 2013 study *Zero-R^2 Hedge Funds and Market Neutrality* and the 2011 paper *Do the Best Hedge Funds Hedge?*[19,20] The former paper showed that about a third of hedge funds had an R^2 that was insignificantly different from zero, which was indicative of the omitted variable issue. The latter study reported an average R^2 of only 0.26 and showed that lower R^2 hedge funds delivered better performance whether measured using Sharpe ratio, information ratio, or alpha.

Forsberg, Gallagher, and Warren combined monthly hedge fund returns from the Hedge Fund Research and eVestment data sets for the period from January 1997 to June 2016 and performed two types of analysis:

- **Performance persistence analysis.**

  The authors used three approaches for evaluating performance persistence: panel regression, Fama-MacBeth regression introduced in the 1973 study *Risk, Return, and Equilibrium: Empirical Tests,* and quartile analysis.[21] As discussed in the aforementioned 2010 paper *Do Hot Hands Exist Among Hedge Fund Managers? An Empirical Evaluation,* the persistence of outperforming top-quartile funds is more important to investors than the persistence of underperforming bottom-quartile funds because they cannot be shorted.[22]

  The quartile analysis was performed by sorting hedge funds based on their cohort alphas using rolling 24 months. The out-of-sample performance was tracked for the following 16 quarters. The authors found that the relative performance of the top-quartile and bottom-quartile hedge funds persisted for up to 12 quarters, and the persistence was stronger when gross returns were used to estimate cohort alphas.

- **Fund-of-funds exercise.**

  The authors further considered practical implications for manager selection by performing a fund-of-funds portfolio analysis. In this exercise, portfolios were formed by equally allocating to the 15 largest cohorts. Within each cohort, its weight was equally allocated across the top two or four funds with the highest cohort alphas resulting in a top-30 or top-60 fund portfolios. Portfolios were rebalanced either quarterly or annually. Cohort alphas were calculated using rolling 24 months excluding the last quarter to account for the delay in hedge fund reporting (this issue was discussed in detail in Section 2.2.1). The benchmark "non-top" portfolios were constructed by allocating equally to the 15 largest cohorts, and within cohorts allocating equally to all funds except the top two or four funds for non-top-30 and non-top-60 portfolios, respectively. The authors found that the top-30 and top-60 outperformed the non-top-30 and non-top-60 portfolios by an economically and statistically significant 1.2 percent to 2.9 percent per annum, respectively.

Appendix B.5 includes detailed implementation steps of the peer cohort alpha approach. It can likely be further improved by sorting funds based on the t-statistic of alpha rather than alphas and using the Newey-West adjustment for the calculation of the t-statistic of alpha rather than a typical OLS regression, as suggested in the aforementioned 2007 paper *Do Hedge Funds Deliver Alpha? A Bayesian and Bootstrap Analysis.*[23].

4.2 INTERESTING NUGGETS: COMBINING QUANTITATIVE AND QUALITATIVE FACTORS WITHIN A BAYESIAN FRAMEWORK

This section shares one interesting nugget: combining quantitative and qualitative factors within a Bayesian framework. As discussed in Section 3.2.6, institutional investors consider both quantitative and qualitative factors. We introduced a Bayesian framework that can be used to combine the two types of factors to detect skilled hedge funds. We demonstrate this framework using the Sharpe ratio, but the methodology can be applied to other performance metrics such as the t-statistic of alpha or information ratio.

Consider a world with two types of hedge funds:

- Skilled funds with a true Sharpe ratio of 1. However, as discussed in this chapter, skilled managers may produce Sharpe ratios that are higher or lower than 1. For example, a skilled hedge fund manager may produce a Sharpe ratio of 0.8 due to bad luck.
- Unskilled funds with a true Sharpe ratio of 0. Unskilled hedge fund managers may produce negative Sharpe ratio or they could produce Sharpe ratios of 0.8 or even 1.5 due to luck.

If we see a hedge fund track record with a Sharpe ratio of 0.8, what is the likelihood that the hedge fund manager is skilled? We can answer this question by estimating the number of skilled hedge funds with a Sharpe ratio of 0.8 and the number of unskilled hedge funds with a Sharpe ratio of 0.8. For example, if we consider a "low quality" pool of hedge fund managers, we have a large number of unskilled managers that have not been eliminated with qualitative due diligence, it may include 18 skilled hedge fund managers with a Sharpe ratio of 0.8 and 162 unskilled hedge fund managers with a Sharpe ratio of 0.8. In this case, the likelihood of observing a skilled hedge fund manager with a Sharpe ratio of 0.8 in that pool is equal to $18/(18+162) = 18/180 = 0.1$ or 10 percent. Thus, although 0.8 Sharpe ratio is much closer to 1 than to 0, the fund is much more likely to be unskilled. This problem is not unique to finance. For example, the New York Times article *Gauging the Odds (and the costs) in Health Screening* by Richard Thaler, a winner of the Nobel Memorial Prize in Economic Sciences, discusses the issue of using mammograms for young women with no risk factors: *Suppose that there is a one-in-1,000 chance that a woman in her 40s with no symptoms has breast cancer, and that 90 percent of the time a mammogram correctly classifies women as having cancer or not. If a woman in this group tests positive on her mammogram, what is the chance that she has cancer? The answer is not 90 percent. It is less than 1 percent, because of the large number of false positive results.*

Now, consider the case of a "high quality" pool of hedge fund managers with a much smaller portion of unskilled hedge fund managers due to an effective qualitative due diligence. For example, if a pool of hedge fund managers includes 18 skilled hedge fund managers with a Sharpe ratio of 0.8 and two unskilled hedge fund managers with a Sharpe ratio of 0.8, the likelihood of observing a skilled hedge fund manager with a Sharpe ratio of 0.8 in that pool is equal to $18/(18+2) = 18/20 = 0.9$ or 90 percent.

Appendix B.6 includes derivation of a threshold Sharpe value for selecting skilled hedge funds given the track record length and a proportion of skilled funds in the hedge fund pool. The role of qualitative due diligence is to increase the proportion of skilled funds in the pool.

Our Bayesian framework demonstrates why sophisticated investors consider both qualitative and quantitative factors in their hedge fund evaluation decisions. Although we used a simple example that relied on Sharpe ratios, the framework can be applied when measuring performance relative to asset pricing models or benchmarks.

4.3 KEY TAKEAWAYS

Following are the key takeaways from this chapter:

- Hedge fund performance evaluation is challenging because of the high role of serendipity. Bootstrapping tests help measure the impact of luck.

- Several interesting performance evaluation approaches include a noise reduced alpha, seemingly unrelated assets, and peer cohort.
- A Bayesian framework is effective for combining quantitative and qualitative factors.

CHAPTER 5

From Mean-Variance to Risk Parity

"It doesn't matter how beautiful your theory is, it doesn't matter how smart you are. If it doesn't agree with experiment, it's wrong."—Richard Feynman.[a]

"It's not whether you're right or wrong that's important, but how much money you make when you're right and how much you lose when you're wrong."—George Soros.[b]

After learning about selecting hedge funds, we turn to another crucial portfolio management topic—portfolio construction. This chapter introduces a practical customizable framework for the evaluation of portfolio construction approaches. We describe the evolution of techniques from mean-variance optimization and its extensions to strategies that diversify risk across managers (risk-parity) and time (volatility-targeting).

5.1 FRAMEWORK FOR PORTFOLIO MANAGEMENT

As discussed in Section 3.1, most academic studies of hedge fund performance persistence are not relevant for institutional investors. The framework of Molyboga, Bilson, and Baek closed the gap between academia and industry by introducing a robust and flexible methodology

[a]Richard Feynman was a Nobel Laureate in physics, 1965.
[b]George Soros is a famous Hungarian-born American investor.

DOI: 10.1201/9781003175209-5

capable of evaluating whether a fund selection technique can within real world constraints benefit a specific institutional investor subject to a unique set of investment objectives and constraints.[1] The 2016 study *A Simulation-Based Methodology for Evaluating Hedge Fund Investments* by Marat Molyboga and Christophe L'Ahelec presented a similar methodology for evaluating portfolio construction approaches that is consistent with investment practices.[2]

5.1.1 A General Framework of Molyboga and L'Ahelec

Molyboga and L'Ahelec introduced a modification of the large-scale simulation framework with real life constraints of Molyboga, Bilson, and Baek. The original framework was used to evaluate persistence in hedge fund managers' performance by ranking funds and then comparing the performance of equally-weighted portfolios of "SKILLED" funds and "ALL" funds. By contrast, Molyboga and L'Ahelec did not attempt to find skilled funds. Instead, they focused on the performance implications of portfolio construction techniques. Their study considered two minimum risk approaches (minimum-variance and minimum semi-standard deviation), three equal risk methods ($1/N$, equal volatility-adjusted, and risk-parity) described in Section 5.2, and a random portfolio approach used as a benchmark. The in-sample/out-of-sample framework mimicked the actions of an institutional investor making allocation decisions at the end of each month. However, the rebalance frequency can be seamlessly adjusted to quarterly, semi-annual, or annual. The study used 10,000 simulations and covered the out-of-sample period between January 1999 and December 2014.

The framework includes the following steps:

- **First step: Data**. The dataset is chosen and adjusted for survivorship and backfill/incubation biases.
- **Second step: Eligible funds for each rebalance period**. For each rebalancing period, the framework excludes all funds that fail to satisfy the investor's investment constraints such as the minimum track record length or AUM. For example, the first decision was made in December 1998. Due to the delay in CTA reporting, the investor had return information only through November 1998. Thus, the investor considered all funds that had a complete set of monthly returns between December 1995 and November 1998. The investor eliminated all funds in the bottom

quintile of AUM among the funds considered because they were too small. This relative AUM threshold was more appropriate than the fixed AUM approach commonly used in the literature because the average level of AUM had increased substantially over the last 20 years.

- **Third step: Single simulation**. Each simulation started at the end of December 1998. The investor randomly selected five funds from the eligible pool of CTAs and allocated to them using the five risk-based approaches and a random portfolio allocation. Monthly returns were recorded for each portfolio construction approach for January 1999 with a liquidation bias adjustment if required. At the end of January 1999, the constituents of the original portfolio that were no longer in the updated eligible pool of funds were randomly replaced with funds from the new pool. Each portfolio was then rebalanced again using the original portfolio construction methodologies.[c] The process was repeated until the end of the out-of-sample period of December 2014. A single simulation resulted in six out-of-sample return streams between January 1999 and December 2014—one for each of the portfolio construction approaches.
- **Fourth step: Large-scale simulation**. Since the methodology produces a large number of feasible portfolio constituents in each period, it relied on a large-scale simulation approach. A large number of simulations were performed to produce multiple time series for each portfolio construction approach. The authors recommended using 10,000 simulations.
- **Fifth step: Performance evaluation of out-of-sample results**. Out-of-sample performance was evaluated using both standalone performance metrics and measures that considered portfolio contribution benefits. Standalone performance metrics included annualized return, *maximum drawdown*, Sharpe ratio, *Calmar ratio*, *Fung-Hsieh alpha*, and t-statistic of alpha. Performance contribution was measured as the resultant difference in the Sharpe ratio and the Calmar ratio from replacing 10 percent of the original portfolio of stocks and bonds with portfolios of CTA

[c]The framework is flexible—the number of funds in a portfolio, rebalancing frequency, AUM threshold levels, and other parameters can be customized to reflect each investor's preferences and constraints.

funds constructed within the simulation framework. Since each performance measure is represented by a distribution that contains 10,000 values, distributions are compared using means and medians for all measures and the percentage of positive values for Fung-Hsieh alpha and the percentage of positive marginal Sharpe and Calmar ratios in the performance contribution measures. Since simulations were not independent, the authors applied a bootstrapping procedure to draw a statistical inference.[d]

As discussed, the authors performed analysis of standalone performance and evaluated the marginal contribution of CTA portfolios to the investor's 60-40 portfolio of stocks and bonds. For brevity, we present only the standalone performance results. Molyboga and L'Ahelec analyzed distributions of out-of-sample returns over the complete data period using means and medians of several performance metrics.[e] For brevity, only the means are presented.

Table 5.1 shows the mean values of the distributions of returns, volatilities, Sharpe and Calmar ratios, and maximum drawdowns for each portfolio construction approach. The superscript star indicates that the performance measure of a given portfolio approach exceeds that of the RANDOM portfolio at the 99 percent confidence level. The subscript star shows that the performance measure of a given portfolio approach is lower than that of the RANDOM portfolio at the 99 percent confidence level.

The minimum risk approaches tended to have the lowest volatilities of the portfolio methodologies considered in the study. MV and MDEV had mean volatilities of around 6.8 percent whereas EVA and RP had volatilities of around 8.21 percent and 8.66 percent, respectively, followed by EN and RANDOM with volatilities that exceed 11 percent. However, the minimum volatility approaches delivered low returns and risk-adjusted returns that were inferior to those of the other approaches. This

[d]The bootstrapping procedure followed each steps of the simulation framework but limited the set of portfolio construction approaches to the Random portfolio methodology to which the authors choose to compare all other approaches. Each simulation set consisted of 10,000 simulations. The bootstrapping procedure included 400 sets of simulations, a sufficient number to estimate p-values with high precision. A comparison of the performance metrics of the original simulation to the bootstrapped sets of simulations gave the p-values reported in the empirical results section.

[e]Since simulations were not independent, the authors used a bootstrapping methodology to draw statistical inferences about the relative performance of portfolio construction approaches.

Table 5.1 **Mean statistics of out-of-sample performance 1999-2014.** This table presents mean values of out-of-sample performance measures for each portfolio construction approach. EN (equal notional or $1/N$), EVA (equal volatility-adjusted, or inverse volatility approach), and RP (risk-parity) are the three equal-risk approaches. MV (minimum variance) and MDEV (minimum semi-standard deviation) are the two minimum-risk approaches. Performance measures include annualized excess return, annualized excess standard deviation, Sharpe ratio, and Calmar ratio (defined as annualized excess return over maximum drawdown). The superscript star indicates that the performance measure of a given portfolio approach exceeds that of the RANDOM portfolio at the 99% confidence level. The subscript star shows that the performance measure of a given portfolio approach is lower than that of the RANDOM portfolio at the 99% confidence level.

Approach	Return	Volatility	Sharpe Ratio	Calmar Ratio
RANDOM	3.72%	11.75%	0.319	0.154
EN	3.73%	$11.03\%_*$	0.342^*	0.168^*
EVA	$2.95\%_*$	$8.21\%_*$	0.358^*	0.174^*
RP	$3.13\%_*$	$8.66\%_*$	0.362^*	0.176^*
MV	$2.13\%_*$	$6.79\%_*$	0.304_*	0.136_*
MDEV	$2.10\%_*$	$6.80\%_*$	0.299_*	0.134_*

finding is consistent with those of the 2009 study *Optimal versus Naive Diversification: How Inefficient is the 1/N Portfolio Strategy?,* which documented the superior out-of-sample performance of the naive 1/N (EN) approach relative to that of several extensions of mean-variance optimization including the minimum variance (MV) approach.[3,f] The three equal-risk approaches had risk-adjusted performance which was superior to that of the RANDOM approach. In contrast, minimum risk approaches yielded inferior results, on average.

While Table 5.1 presented the mean values of several performance metrics, a complete evaluation of the portfolio construction methodologies was also considered in the study. Molyboga and L'Ahelec concluded that the equal-risk approaches were superior to the minimum-risk approaches.

[f] Jensen's inequality suggests the EN approach should dominate the RANDOM methodology in terms of Sharpe ratio due to the concavity of the Sharpe ratio.

5.1.2 Mean-Variance Optimization: A Beautiful Theory with Ugly Results

Portfolio selection has been a fruitful area of research since the introduction of the *parsimonious theory*, proposed in the 1952 study *Portfolio Selection* by Harry Markowitz, which reduced a complex *asset allocation* problem to a simple calculation that relies solely on the vector of expected returns and the covariance matrix.[4] Unfortunately, as discussed in the 1989 study *The Markowitz Optimization Enigma: Is Optimized Optimal?* and the 1991 paper *On the Sensitivity of Mean-Variance-Efficient Portfolios to Changes in Asset Means: Some Analytical and Computational Results,* this beautiful theory produces ugly results as evidenced by poor out-of-sample performance and extreme, unstable portfolio weights.[5,6]

The issue of instability can be visualized by considering a simple example with four assets A, B, C, and D. Table 5.2 displays the annualized volatilities and pair-wise correlations of the four assets. Assets A and B are highly correlated similarly to the assets C and D.

Table 5.2 **Mean-variance optimization example: volatilities and pair-wise correlations of the four assets.** This table shows annualized volatility and pair-wise correlations of the four assets.

Assets	A	B	C	D
Volatility	12%	12%	15%	15%
Correlation Matrix	1	0.8	0.6	0.3
		1	0.3	0.3
			1	0.7
				1

Table 5.3 shows three very similar sets of assumptions regarding expected returns and the corresponding mean-variance optimal portfolios.[7] The expected returns of assets A, B, and D are fixed across the scenarios and equal to 12%, 12%, and 15%, respectively. The expected return of asset C varies between 14% and 16%, a very small change, particularly given the challenge of estimating expected returns highlighted by Robert Merton in the 1980 study *On Estimating the Expected Return on the Markets: An Exploratory Investigation.*[8] In fact, if we asked 20 people about their expectation regarding next year's stock market return, it wouldn't be surprising to see a wide range of

expectations that includes both −15 percent and +15 percent. The optimal weights vary substantially across the scenarios. The optimal allocation to C varies between 7 percent and 41 percent, almost 35 percentage points. The optimal allocation to A ranges between −16 percent and 22 percent, nearly 40 percentage points.

Table 5.3 **Mean-variance optimization example: expected returns and mean-variance optimal portfolio weights.** This table shows three sets of assumptions regarding expected returns of the four assets and corresponding optimal portfolio weights.

Assets	A	B	C	D
Expected Returns	12%	12%	15%	15%
Optimal weights	3%	47%	24%	26%
Expected Returns	12%	12%	16%	15%
Optimal weights	−16%	60%	41%	16%
Expected Returns	12%	12%	14%	15%
Optimal weights	22%	35%	7%	37%

The issue of instability is driven by the high sensitivity of the Markowitz portfolio weights to *estimation error.*[g,9,10] This difficulty is further compounded by the problem of estimating the vector of expected returns with a high degree of precision, as noted by Robert Merton in the 1980 study *On Estimating the Expected Return on the Markets: An Exploratory Investigation.*[11]

5.1.3 Extensions of Mean-variance Optimization

In response to the issues of instability and poor out-of-sample performance, several extensions of the mean-variance optimization have

[g]Markowitz portfolios are obtained by solving a quadratic problem, which requires the inversion of a covariance matrix. The 2012 study *Balanced Baskets: A New Approach to Trading and Hedging Risks* and the 2016 paper *A New Diagnostic Approach to Evaluating the Stability of Optimal Portfolios* showed that the magnitude of the sensitivity issue could be assessed using the condition number of the covariance matrix. If the covariance matrix is near singular, the condition number, defined as the ratio of the largest to the smallest eigenvalues of the covariance matrix, is large and the portfolio weights are highly sensitive to estimation error.

emerged. The 2009 study *Optimal versus Naive Diversification: How Inefficient is the 1/N Portfolio Strategy?* by Victor DeMiguel, Lorenzo Garlappi, and Raman Uppal considered 14 versions of mean-variance optimization that included Bayesian approaches to estimation error, moment restrictions, portfolio constraints, and optimal combinations of portfolios.[12] Although the authors showed that none of the approaches considered in the study could outperform a simple 1/N approach out-of-sample, we want to highlight a few popular extensions of mean-variance optimization:

- Mean-variance with shrinkage,
- Black-Litterman optimization,
- Minimum variance portfolio.

Following is a brief review of the three approaches.

5.1.3.1 Mean-variance with Shrinkage

One way to improve the performance of the classic Markowitz mean-variance optimization is to acknowledge that sample means and sample covariance matrices are poor forward-looking estimates and try to improve the quality of estimation by applying Bayesian techniques or shrinkage estimators.

In the 1956 study *Inadmissibility of the Usual Estimator for the Mean of a Multivariate Normal Distribution* Charles Stein showed that a sample mean was a poor estimator for the mean of a multivariate normal distribution when a quadratic loss function was considered.[13] Instead, Stein proposed a new type of an estimator, which is a weighted average of the sample mean and a target value θ_0, which can be any vector of the same size.

Stein also showed that a shrinkage estimator can be superior to the sample mean estimator for any target value θ_0. Gains can be higher if the target value and the weight given to the sample mean are chosen well. Thus, since shrinkage estimators can improve the accuracy of the mean and the covariance matrix, they can improve the performance of mean-variance optimization.

The 1986 study *Bayes-Stein Estimation for Portfolio Analysis* by Philippe Jorion introduced a shrinkage estimator of the mean and showed that it improved the performance of mean-variance optimization.[14]

In their 2003 study *Improved Estimation of the Covariance Matrix of Stock Returns with an Application to Portfolio Selection* and their 2004 study *Honey, I Shrunk the Sample Covariance Matrix* Olivier Ledoit and Michael Wolf applied a similar shrinkage approach to improve the estimation quality for the covariance matrix.[15,16] Ledoit and Wolf stated that "no one should use the sample covariance matrix for portfolio optimization" because shrinkage estimation systematically reduced estimation error when it mattered most.

Appendix C.1 presents a detailed overview and calculations of popular shrinkage estimators of means and covariance matrices that can be used as inputs in mean-variance optimization.

5.1.3.2 Black-Litterman Optimization

Black-Litterman optimization introduced in the 1992 study *Global Portfolio Optimization* by Fischer Black and Robert Litterman contributed to the field of quantitative portfolio management by elegantly applying Bayesian statistics to combine two seemingly contradictory ideas: the efficiency of the market portfolio and the benefit of expert opinions that may discover inefficiencies that are hidden to the market participants.[17]

Black-Litterman optimization includes four steps:

1. Capitalization-weighted market portfolio is a good starting portfolio for any investor as it is the optimal equilibrium portfolio according to the Capital Asset Pricing Model as shown in the 1964 study *Capital Asset Prices: A Theory of Market Equilibrium Under Conditions of Risk* by William Sharpe.[18]

2. Reverse optimization produces the implied expectations of assets' performance.

3. Expert opinions are defined in terms of expectations about absolute or relative performance of assets with a certain degree of confidence/uncertainty.

4. Black-Litterman framework incorporates the expert opinions and produces a new set of portfolio weights.

Black-Litterman optimization solves the instability problem discussed in Section 5.1.2 and results in portfolios that make intuitive sense. For example, if an investor has no private views about expected returns,

Black-Litterman optimization produces a market portfolio—a good option. If the investor has an opinion about a small number of assets based on fundamental or quantitative analysis, given his confidence in the opinion the framework can incorporate it and produce a well-diversified portfolio—one, which is expected to outperform the market portfolio if the investor is correct. Thus, Black-Litterman optimization is a popular portfolio technique used by practitioners. Appendix C.2 provides a detailed description of each step with formulas and examples.

5.1.3.3 Minimum-variance Portfolio

Another approach to overcoming poor out-of-sample performance and instability of portfolio weights of mean-variance optimization was introduced in the 1992 study *When Will Mean-Variance Efficient Portfolios Be Well Diversified?* and the 2003 paper *Risk Reduction in Large Portfolios: Why Imposing the Wrong Constraints Helps*—using minimum-variance portfolios and imposing constraints of non-negative weights.[19,20]

The former study argued that the extreme positive and negative weights observed in mean-variance portfolios were symptoms of strong factor structure in the assets covariance matrix and proposed the minimum variance portfolio as a way to mitigate the effect of estimation error in the mean on portfolio weights. The latter study argued that the estimation error in the sample mean was so large that little was lost by ignoring the mean altogether and investigated the impact of non-negativity constraints on out-of-sample performance. They found:

- Imposing non-negativity constraints on portfolio weights of minimum variance and minimum tracking error portfolios based on the sample covariance matrix improved their out-of-sample performance almost as much as estimators that relied on factor models, shrinkage estimators and daily returns.

- Using daily returns for estimation of covariance matrices led to the best out-of-sample performance among unconstrained minimum variance and minimum tracking error portfolios and corrections for the microstructure effects provided no additional benefit.

- Minimum variance portfolios outperformed mean-variance portfolios regardless of constraints suggesting that the estimates of the mean returns were too noisy to be useful.

5.2 FROM MEAN-VARIANCE TO RISK PARITY

While many papers have attempted to reduce the impact of estimation error, the influential 2009 study *Optimal versus Naive Diversification: How Inefficient is the 1/N Portfolio Strategy?* by Victor DeMiguel, Lorenzo Garlappi, and Raman Uppal showed that even very sophisticated approaches failed to outperform a naive 1/N portfolio and argued that the estimation window required to capture the potential gains of optimal portfolios was too long.[21] The failure of mean-variance optimization led to the popular risk-parity approach discussed in detail in the 2010 paper *The Properties of Equally Weighted Risk Contribution Portfolios.*[22] The risk-parity approach ignores return forecasts and instead attempts to maximize diversification by allocating risk equally across portfolio constituents.

5.2.1 How Inefficient Is $1/N$?

The aforementioned 2009 paper *Optimal versus Naive Diversification: How Inefficient is the 1/N Portfolio Strategy?* investigated the conditions under which mean-variance optimal portfolios would perform well even in the presence of estimation risk.[23] Following the advice of Rabbi Issac bar Aha who suggested in the fourth century "One should always divide his wealth into three parts: a third in land, a third in merchandise, and a third ready to hand," as a benchmark they used a naive $1/N$ portfolio that gave equal allocation to each asset considered for allocation.

In addition to the naive $1/N$ approach, the study considered 14 models that represented five categories of mean-variance optimization:

- **Classical approach that ignores estimation error**: sample-based mean-variance optimization.
- **Bayesian approach to estimation error**: Bayesian diffuse-prior, Bayes-Stein, and Bayesian data-and-model.
- **Moment restrictions**: minimum-variance, value-weighted market portfolio, and the missing factor model from the 2000 paper *Asset Pricing Models: Implications for Expected Returns and Portfolio Selection* by Craig MacKinlay and Lubos Pastor.[24]

- **Portfolio constraints**: sample-based mean-variance with shortsale constraints, Bayes-Stein with short-sale constraints, minimum-variance with short-sale constraints, and minimum-variance with generalized constraints.

- **Optimal combination of portfolios**: the "three-fund" model from the 2007 paper *Optimal Portfolio Choice with Parameter Uncertainty,* a mixture of minimum-variance and $1/N$, and the multi-prior model from the 2007 paper *Portfolio Selection with Parameter and Model Uncertainty: A Multi-Prior Approach.*[25,26]

The authors used the standard performance measures of Sharpe ratios, certainty-equivalent return (CER), and turnover to investigate the out-of-sample performance of the portfolio management techniques. The study used seven empirical equity-specific datasets with monthly returns that included:

1. Ten sector portfolios of the S&P 500 and the U.S. equity market portfolio.

2. Ten industry portfolios and the U.S. equity market portfolio.

3. Eight country indices and the World Index.

4. SMB (size) and HML (value) portfolios and the U.S. equity market portfolio.

5. Twenty size and book-to-market (value) portfolios and the U.S. equity market portfolio.

6. Twenty size and book-to-market portfolios and the U.S. equity market, SMB and HML portfolios.

7. Twenty size and book-to-market portfolios and the U.S. equity market, SMB, HML, and UMD (momentum) portfolios.

While the constrained minimum-variance approach introduced in the aforementioned 2003 study *Risk Reduction in Large Portfolios: Why Imposing the Wrong Constraints Helps* produced the best results relative to the other versions of mean-variance portfolios, it failed to outperform the naive 1/N approach.[27]

In order to understand the conditions under which mean-variance optimal portfolios would perform well even in the presence of estimation risk, DeMiguel, Garlappi, and Uppal provided an analytic expression for the critical length of the estimation window that was required for the classic sample-based mean-variance strategy to outperform the 1/N approach.

They found that the estimation window was a function of three variables: the number of assets, the true ex-ante Sharpe ratio of the mean-variance efficient portfolio, and the Sharpe ratio of the 1/N portfolio.
The estimation window had:

- A positive relation with the number of assets. Since a bigger number of assets requires estimating a bigger number of parameters, the estimation error is higher, and, therefore, a longer window is required to sufficiently reduce the estimation error.
- A negative relation with the true ex-ante Sharpe ratio of the mean-variance efficient portfolio.
- A positive relation with the Sharpe ratio of the 1/N portfolio.

When calibrating the model to U.S. stock-market data, DeMiguel, Garlappi, and Uppal found that the critical window was 3,000 months for a portfolio with 25 assets and more than 6,000 for a portfolio with 60 assets. This finding questioned the common practice of using 60–120 month windows in portfolio optimization. Simulation-based analysis showed that the other extensions of mean-variance optimization also required very long estimation windows to outperform the naive 1/N strategy.

5.2.2 Naive $1/N$, Minimum-variance, or Equal Risk?

While DeMiguel, Garlappi, and Uppal made a compelling case for the 1/N approach as an alternative to more sophisticated optimization approaches, it has been criticized in the literature.[28] For example, the 2010 study *In Defense of Optimization: The Fallacy of* $1/N$ argued that mean-variance optimization could be effective if it relied on forward-looking inputs that were based on economic intuition rather than backward-looking inputs that were estimated using realized returns.[29]

Risk-parity, or equal risk, is another portfolio construction approach that attempts to improve the performance by focusing on diversification of risk across portfolio constituents. While equal risk approaches are discussed in details in Section 5.2.3, this section uses a simple hypothetical example with two uncorrelated assets A and B to compare the naive 1/N, minimum-variance and an equal risk approach represented by an equal volatility-adjusted (EVA) approach highlighted in the 2012 study *A Proof of Optimality of Volatility Weighting over Time* and discussed in the 2016 paper *A Simulation-Based Methodology for Evaluating Hedge Fund Investments.*[30,31] The EVA approach produces weights that are inversely related to assets' volatilities in an attempt to equally balance the risk contribution from each asset. For example, if the volatility of asset A is equal to 10 percent and the volatility of asset B is equal to 20 percent, the EVA approach allocates 2/3 to asset A, which is twice as much as the 1/3 allocation to asset B. In this case, the risk contribution from asset A is equal 10 percent∗2/3, which is identical to 20 percent∗1/3, the risk contribution from asset B.

Appendix C.4 provides detailed derivations and Table 5.4 summarizes the conditions under which the 1/N, minimum-variance, or equal-risk approaches produce the highest Sharpe portfolio.

Table 5.4 **A hypothetical example with 1/N, minimum-variance and equal-risk approaches.** This table describes the conditions under which the portfolio approaches produce the highest Sharpe portfolio and the expected excess return of asset B given the assumptions of the volatility of asset A equal to 10 percent, the volatility of asset B equal to 20 percent, the expected return of asset A equal to 5 percent, and no correlation between assets A and B.

	Condition	μ_B
1/N	Expected excess returns are proportional to variances	20%
Minimum-variance	Expected excess returns are the same across assets	5%
Equal-risk	Sharpe ratios are the same across assets	10%

5.2.3 Standard Risk Parity: Equal Risk Contribution

In this section, we consider the classical risk parity approach discussed in the 2006 study *On the Financial Interpretation of Risk Contribution: Risk Budgets Do Add Up,* the 2010 paper *The Properties of Equally Weighted Risk Contribution Portfolios,* the 2013 paper *Risk Parity, Maximum Diversification, and Minimum Variance: An Analytic*

Perspective, and the 2013 study *Are Risk-Parity Managers at Risk Parity?*.[32,33,34,35] It is different from the EVA approach from the 2012 study *A Proof of Optimality of Volatility Weighting over Time* that only considers volatilities of the assets because risk parity also incorporates correlations to allocate risk equally across assets.[36]

Since this topic is highly technical, Appendix C.5 is extensive and covers:

- A detailed discussion and derivations of several important terms such as the marginal risk contribution, the total risk contribution, and the percentage risk contribution. If all pair-wise correlations are equal to one another, the equal risk contribution approach produces equal volatility-adjusted allocations.

- A discussion of a generalized risk parity approach. We show that the risk parity idea can be applied to popular risk measures of expected shortfall, also known as conditional value-at-risk (CVaR) or expected tail loss, maximum drawdowns or maximum loss. The expected shortfall is probably the most interesting example because it is highly regarded by regulators and practitioners as one of the best measures of risk. It is defined as the expected (average) loss beyond the VaR level.

- A detailed discussion of a risk-parity approach with modified conditional expected drawdown introduced in the 2017 study *Portfolio Management with Drawdown-Based Measures*.[37]

  Drawdown-based analysis is important because best practices in due diligence of alternative investments require drawdown analysis as part of standard quantitative due diligence.[38]

5.2.4 Adaptive Optimal Risk Budgeting

The adaptive optimal risk budgeting (AORB) approach that is based on risk contribution, but attempts to produce an approximately mean-variance efficient solution when Sharpe ratios and correlations vary across assets and time, was introduced in the 2020 study *Adaptive Optimal Risk Budgeting*.[39] The method was designed to overcome two key weaknesses of the classical risk parity approach:

- Risk parity does not account for the variability of Sharpe ratios and correlations across assets.

- Risk parity ignores historical data that could be useful for predicting Sharpe ratios.

The AORB approach is based on the finding from the 2001 paper *Implementing Optimal Risk Budgeting* and the 2006 study *The Sense and Nonsense of Risk Budgeting* that showed that the optimal mean-variance risk budget vector could be expressed as a function of the vector of Sharpe ratios and the correlation matrix.[40,41]

If the Sharpe ratios and the correlations are the same for all portfolio constituents, the expression produces the classic risk-parity portfolio. If the Sharpe ratios and the correlations are estimated using historical data, the expression results in the classic mean-variance solution.

The AORB approach allows for the Sharpe ratios to vary across portfolio constituents and across time. The AORB initially assumes that the Sharpe ratios are the same for all portfolio constituents but then learns from the historical data and adjusts the Sharpe ratios of the assets. Appendix C.6 covers the technical details of the AORB approach.

The authors used simulated data to evaluate the method and found that the AORB approach outperformed risk-parity under a broad set of conditions. They argued that the approach was relevant for portfolios of risk premia because factors were often associated with evolving correlation structure of returns and Sharpe ratios that gradually degraded, as documented in the 2019 paper *Alice's Adventures in Factorland: Three Blunders That Plague Factor Investing.*[42] Since hedge funds exhibit similar characteristics, the AORB approach may also be effective for hedge fund portfolios.

The 2021 study *Fuzzy Factors and Asset Allocation* extended the AORB methodology by applying a fuzzy set theory to deal with the vagueness of investment objectives, time-varying characteristics of portfolio constituents and their links to risk factors.[43] The authors argued that their "fuzzy" asset allocation approach was particularly relevant when used for custom strategic asset allocation solutions.

5.2.5 Diversification Across Time with Volatility Targeting

As we have shown, hedge fund investors can benefit from portfolio construction approaches such as risk parity that attempt to diversify risk across hedge funds. However, relatively little work is dedicated to exploring diversification across another dimension—the dimension of time—with portfolio volatility targeting. Volatility targeting dynamically scales aggregate portfolio leverage to achieve constant expected portfolio volatility and, thus, allocate risk equally across time.

This lack of research is surprising since a large number of academic papers have reported the performance benefits of volatility targeting for risk premia strategies. For example, the 2015 study *Momentum Has Its Moments* and the 2016 paper *Momentum Crashes* showed that volatility targeting, or adjusting exposure to target a constant ex-ante volatility, nearly doubled the Sharpe ratio of cross-sectional momentum.[44,45] The 2016 study *Time Series Momentum and Volatility Scaling* reported that the abnormal returns of time-series momentum were largely driven by volatility scaling.[46] The 2020 paper *Short-Term Trend: A Jewel Hidden in Daily Returns* showed that volatility scaling improved performance of time-series momentum strategies across asset classes and parameter sets.[47]

The 2017 study *Volatility-Managed Portfolios* extended evidence for benefits of volatility scaling to a broad array of risk premia such as market, value, currency carry, and betting-against-beta.[48] The 2020 study *On the Performance of Volatility-Managed Portfolios* challenged that finding due to a potential methodological issue.[49,50] The authors suggested using out-of-sample Sharpe ratios for performance evaluation and found mixed evidence of benefits of volatility targeting. Volatility scaling improved performance of some strategies such as momentum-based strategies, profitability, and betting-against-beta, but failed to yield statistically positive results for most other strategies considered in the study.

Academic literature has provided little evidence regarding the impact of volatility targeting on active strategies such as portfolios of mutual funds or hedge funds. One exception is the 2021 study *Should Mutual Fund Investors Time Volatility?* that reported that volatility-targeting mutual funds produced significantly higher alphas and Sharpe ratios.[51] The authors also found that the performance improvement was driven by both volatility timing and return timing.

Another exception is the 2019 paper *Portfolio Management of Commodity Trading Advisors with Volatility Targeting* by Marat Molyboga. Molyboga empirically investigated the impact of volatility targeting on multi-CTA portfolios within the large-scale simulation framework of Molyboga and L'Ahelec discussed in Section 5.1.1.[52] Molyboga also derived conditions that should be satisfied for volatility targeting to improve the out-of-sample Sharpe ratio of a hedge fund portfolio.

Appendix C.7 summarizes the theoretical results regarding conditions that should be satisfied for volatility targeting to improve the out-of-sample Sharpe ratio of a hedge fund portfolio. The overall conclusion is that volatility targeting was generally expected to outperform except under a very strict set of conditions of a very strong positive relationship between volatility and expected Sharpe ratios. Molyboga concluded that the impact of volatility targeting could vary across hedge fund strategies and proposed the inequality (C.64), shown in Appendix C.7, that could serve as a rough diagnostic test to evaluate the potential impact. Moreover, he suggested that the large-scale simulation framework of Molyboga and L'Ahelec could be used to evaluate the strategy given real life constraints.

Molyboga imposed the Molyboga and L'Ahelec framework on multi-CTA portfolios with estimates of covariance matrices that were based on a combination of exponential weighting and Ledoit-Wolf shrinkage discussed in Section 5.1.3.1. Exponential weighting is often used in risk management and portfolio management to capture the heteroskedasticity of financial returns in estimation of covariance matrices and variances. The *exponentially weighted moving average* (EWMA) approach is used by RiskMetrics, a highly regarded provider of risk analytics.[53] It is also closely linked to the famous *GARCH* model.[54] Moreover, the 2004 paper *Exponential Weighting and Random-Matrix-Theory-Based Filtering of Financial Covariance Matrices for Portfolio Optimization* recommended applying exponentially weighted covariance matrices to portfolio optimization problems.[55] The estimation was further enhanced by applying the *Ledoit-Wolf shrinkage*

Molyboga found that:

- Volatility targeting improved the out-of-sample returns between 0.53 percent and 0.80 percent per annum, on average.
- The performance improvement grew with the number of managers in the portfolio and yielded positive results in 70 percent to 95 percent of simulations depending on the portfolio size and the portfolio construction methodology considered.

5.3 KEY TAKEAWAYS

Following are the key takeaways from this chapter:

- Hedge fund portfolio construction should be customized to the specific objectives and constraints of individual investors.

- Mean-variance optimization is a beautiful theory with ugly results. Its extensions fail to outperform a naive $1/N$ approach.
- Hedge fund investors can improve performance by diversifying risk across strategies (risk-parity) and time (volatility-targeting).

CHAPTER 6

Advanced Portfolio Construction

"There is no reason and no way that a human mind can keep up with an artificial intelligence machine by 2035."—Gray Scott.[a]

"Innovation distinguishes between a leader and a follower."—Steve Jobs.[b]

While conventional equal-risk approaches provide a solid foundation, investors can further improve on the efficiency of their portfolios. This chapter describes advanced portfolio construction techniques that are relevant for both hedge fund investors and managers. It covers several *machine learning* approaches, two recent cutting-edge methods, and several interesting complementary nuggets.

6.1 PORTFOLIO MANAGEMENT WITH MACHINE LEARNING

In this section, we consider several portfolio management techniques inspired by the machine learning literature.

6.1.1 HRP: Hierarchical Risk Parity

As discussed in Marcos Lopez de Prado's 2016 paper *Building Diversified Portfolios that Outperform Out of Sample* most bottom-up approaches to portfolio construction, such as mean-variance optimization or risk

[a]Gray Scott is an expert in emerging technology.
[b]Steve Jobs is regarded as a pioneer of the personal computer revolution.

DOI: 10.1201/9781003175209-6

parity, tend to rely heavily on each value of a correlation matrix, implicitly implying that any two securities are potential substitutes for one another.[1] This assumption is inconsistent with the practices of investment professionals, particularly in light of the influential 1995 paper *Determinants of Portfolio Performance* by Gary Brinson, Randolph Hood, and Gilbert Beebower that highlighted the importance of top-down asset allocation decisions.[2] If an investor attempts to build a 60-40 portfolio of stocks and bonds—a typical starting point for a U.S. institutional investor according to *The Evolution of Equity Mandates in Institutional Portfolios*—the investor would group stocks and bonds separately rather than consider individual stocks and bonds as substitutes for each other.[3] The aforementioned 2016 paper *Building Diversified Portfolios that Outperform Out of Sample* showed that considering a *hierarchical tree structure* allowed for focusing on a small number of relations that were consistent with the top-down perspective of institutional investors who built portfolios by starting at the asset class level and then going down to the level of the individual securities.[4]

Table 6.1 summarizes the relative degree of robustness of four allocation methodologies of 1/N, EVA, standard risk parity (equal risk contribution), and mean-variance optimization to estimation error. The risk parity and mean-variance optimization are the most sensitive because of the excessive reliance on each value of a correlation matrix.

Table 6.1 **Robustness of bottom-up approaches to estimation error.** This table shows which inputs should be estimated for the four bottom-up allocation approaches: the naive 1/N, the equally volatility-adjusted (EVA), the equal risk contribution (ERC), and the mean-variance optimization. It also shows the relative degree of their robustness to estimation error.

	1/N	EVA	ERC	Mean-Variance
Volatility	No	Yes	Yes	Yes
Correlations	No	No	Yes	Yes
Return	No	No	No	Yes
Robustness	Highest	High	Low	Lowest

As we demonstrate later, hierarchical trees help highlight potential weaknesses of the equal risk approaches: EVA and ERC. EVA ignores any links among assets while ERC considers that any two assets are potential substitutes.

EVA completely ignores any potential links among assets, potentially leading to concentrated portfolios. We illustrate this issue by considering a simple example with four assets A, B, C and D, where the assets are mostly uncorrelated except assets A and B have a pair-wise correlation of 0.9. We assume that all four assets have the same volatility of 12 percent. Table 6.2 displays the pair-wise correlations of the four assets. Intuitively, we know that the portfolio is exposed to three uncorrelated factors: the first one is closely correlated to A and B, and the other two are linked to C and D. Thus, building a well diversified portfolio would involve giving roughly equal allocation to each of those factors. However, since EVA ignores any potential links among assets, it allocates about 50 percent to the first factor and 25 percent to each of the remaining factors resulting in a concentrated portfolio that is over-allocated to the first factor. In this case, the ERC approach is more effective at producing a well-diversified portfolio as it allocates 20 percent to A and B, and 30 percent to C and D.

Table 6.2 **Equal volatility-adjusted allocation example.** This table shows the pair-wise correlations of the four asset.

Assets	A	B	C	D
A	1	0.9	0	0
B		1	0	0
C			1	0
D				1

Figure 6.1 shows the hierarchical tree for the correlation matrix from Table 6.2. Assets A and B are close substitutes but they are very different from the assets C and D, which are uncorrelated. The tree makes it clear that A and B share a common factor and, thus, should share an allocation that is given to that factor, whereas C and D are distinct factors.

ERC has the different problem of considering any two assets as potential substitutes for each other, leading to portfolios that are not optimally diversified. We illustrate this issue by considering the simple example from Section 5.1.2 with four assets A, B, C, and D. Table 6.3 displays the pair-wise correlations of the four assets. We can see from the correlation matrix that there are two sets of asset pairs that are similar:

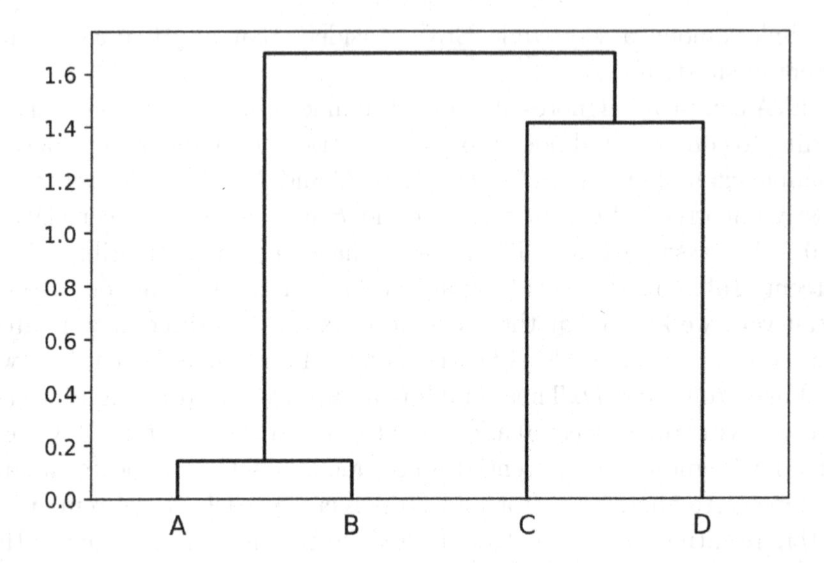

Figure 6.1 Hierarchical tree for the correlation matrix from Table 6.2.

A, B, C, and D. Following a top-down approach we would want to determine how to allocate between the two sets and then give a roughly equal allocation within each set. However, the equal risk contribution approach will result in allocations that are very sensitive to the value of the correlation between A and C.

Table 6.3 **Equal risk contribution methodology example.** This table shows the pair-wise correlations of the four assets.

Assets	A	B	C	D
A	1	0.8	0.6	0.3
B		1	0.3	0.3
C			1	0.7
D				1

Figure 6.2 shows the hierarchical tree for the correlation matrix from Table 6.2. The tree has two clusters—the first includes two similar assets A and B, and the second includes similar assets C and D.

The tree structure helps visualize a top-down hierarchy that is relevant for institutional investors. The HRP approach of Marcos Lopez de Prado applied *graph theory* and machine learning techniques to

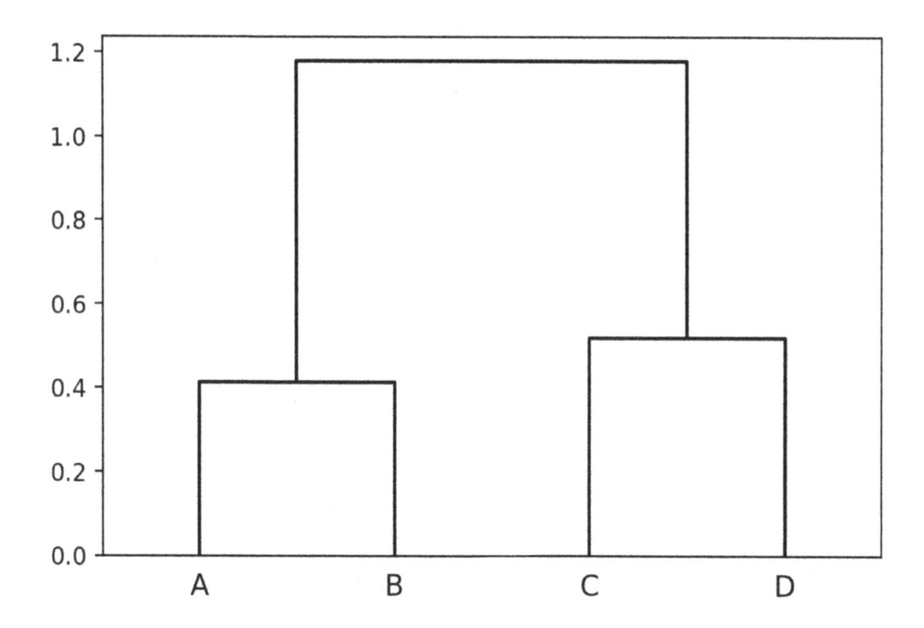

Figure 6.2 Hierarchical tree for the correlation matrix from Table 6.3.

build well-diversified portfolios that performed well out-of-sample.[5] The methodology relies on the top-down tree structure and follows three steps:

1. Tree clustering. Hierarchical clustering is performed using a sample correlation matrix. This step highlights important top-down relations among individual portfolio constituents and their clusters.

2. Quasi-diagonalization. Similar investments are placed together and dissimilar investments are placed far apart. This procedure reshuffles rows and columns of the original correlation matrix so that the largest values are close to the diagonal and the smallest values are away from the diagonal.

3. Recursive bisection. This step allocates across and within clusters using an inverse variance allocation (henceforth, IVA) approach.[6]

Lopez de Prado used a random dataset with 10 assets to compare diversification characteristics of the HRP approach to those of the minimum-variance and the inverse-variance approaches. He found that the *HRP* approach resulted in better diversified portfolios than the minimum-variance approach. For example, the top five holdings represented 92.66 percent weight for the minimum-variance approach and only 62.57 percent for the *HRP* approach. Moreover, the minimum-risk approach gave zero weight to three assets. Lopez de Prado showed that the HRP approach provided a compromise between a highly concentrated minimum-variance approach and the risk-parity-like inverse variance approach.

Lopez de Prado also performed out-of-sample Monte Carlo simulations and found that although the minimum-variance approach produced the lowest variance in-sample, the *HRP* approach produced the lowest variance out-of-sample when compared to the other two approaches. That led him to conclude that the *HRP* approach delivered well-diversified portfolios that outperformed out-of-sample.

6.1.2 MHRP: Modified Hierarchical Risk Parity

The 2020 study *A Modified Hierarchical Risk Parity Framework for Portfolio Management* enhanced the *HRP* approach by adding three intuitive elements commonly used by practitioners.[7] This modified approach:

- Replaced the sample covariance matrix with an exponentially weighted covariance matrix with Ledoit-Wolf shrinkage introduced in the 2019 paper *Portfolio Management of Commodity Trading Advisors with Volatility Targeting.*[8]
- Improved diversification across portfolio constituents both within and across clusters by relying on an equal volatility allocation, EVA, approach rather than an IVA approach, as suggested in the 2012 study *A Proof of Optimality of Volatility Weighting over Time.*[9]
- Improved diversification across time by applying volatility targeting to portfolios as discussed in Section 5.2.5.

In his 2016 study *Building Diversified Portfolios that Outperform Out of Sample* Lopez de Prado suggested that the IVA was optimal when the covariance matrix was diagonal because it produced the minimum variance portfolio.[10] However, as discussed in Section 5.2.2, minimum risk portfolios perform well when expected excess returns are the same across portfolio constituents. In contrast, the EVA approach produces superior Sharpe ratio if the portfolio constituents have roughly the same Sharpe ratios—a more reasonable assumption for hedge funds.

The just cited 2020 study *A Modified Hierarchical Risk Parity Framework for Portfolio Management* imposed the large-scale simulation framework of Molyboga and L'Ahelec on a BarclayHedge CTA sample of 528 live and 1,113 defunct funds over the period 2002–2016.[11] The author found that each enhancement improved out-of-sample Sharpe ratios of multi-CTA portfolios by 13 percent to 19 percent, on average. Moreover, when all three enhancements were combined into a unified *MHRP* approach, they yielded a striking improvement in the out-of-sample Sharpe ratio of 50 percent, on average, with a meaningful reduction in downside risk.

The author also argued that the *MHRP* approach combined the structural benefits of the *HRP* approach with the practical ideas of improved covariance matrix estimation and diversification across time and portfolio constituents. Thus, the *MHRP* approach can be a potentially attractive portfolio management technique for institutional investors.

6.1.3 Beyond MHRP—Denoising Correlation Matrices

Most portfolio construction approaches rely on correlation matrices. Correlations are also important for risk management because large portfolio losses are often driven by the correlated moves of their constituents. Since estimation of correlation matrices is performed with limited amount of data (i.e., the track record length of hedge funds is the same order of magnitude as the number of hedge funds in a portfolio), estimated correlation matrices are noisy, leading to substantial estimation errors. Section 5.1.3.1 describes several shrinkage approaches.[12,13,14] An alternative approach to cleaning correlation matrices is rooted in *random matrix theory.* The 2017 study *Cleaning Large Correlation Matrices: Tools from Random Matrix Theory* provides a comprehensive overview of the random matrix theory and its applications to the problem of cleaning correlation matrices.[15]

In their 2016 paper *Cleaning Correlation Matrices,* Joel Bun, Jean-Philippe Bouchaud, and Marc Potters discussed four standard cleaning approaches (basic linear shrinkage, advanced linear shrinkage, *eigenvalue* clipping, and eigenvalue substitution) and proposed a new approach: *rotationally invariant, optimal shrinkage.*[16] The paper compared their performance and concluded that the new approach represented "a new cleaning recipe that outperforms all existing estimators in terms of the out-of-sample risk of synthetic portfolios." Thus, they recommended using the *rotationally invariant estimators* for large correlation matrices. Appendix D.1 includes technical details of the five approaches covered in the paper.

6.2 CUTTING EDGE PORTFOLIO MANAGEMENT APPROACHES

This section discusses two cutting-edge approaches: the maximum-Sharpe-ratio estimated and *sparse regression approach* introduced in the 2018 study *Approaching Mean-Variance Efficiency for Large Portfolios* and the robust-mean-variance approach from the 2021 paper *Robust Portfolio Choice.*[17,18]

6.2.1 Mean-variance Efficiency for Large Portfolios

The 2018 study *Approaching Mean-Variance Efficiency for Large Portfolios* proposed the maximum-Sharpe-ratio estimated and sparse regression ("MAXSER") approach that was designed for portfolios with a large number of assets such as hedge fund portfolios which can contain hundreds or thousands positions.[19] The authors showed that the MAXSER method could accomplish two objectives simultaneously:

1. Achieve mean-variance efficiency,
2. Satisfy the risk constraint.

The authors performed simulation analysis with parameters for generating returns calibrated from the S&P 500 Index constituents and empirical analysis using monthly returns of the constituents of the Dow Jones 30 Index and the S&P 500 Index. They compared the MAXSER approach to the three-fund Kan and Zhou portfolio from the 2007 paper *Optimal Portfolio Choice with Parameter Uncertainty* and 12 versions of the mean-variance and global minimum variance portfolios.[20] The simulation analysis showed that the MAXSER approach was more accurate at estimating future volatility and produced the highest Sharpe ratio when compared to all portfolio construction approaches considered in the study. When using the lookback of 240 months for parameter estimation, MAXSER achieved approximately 76 percent of the theoretical maximum Sharpe ratio whereas the second-best approach attained 64 percent.

Empirical analysis with the Dow Jones 30 Index constituents for the period between 1977 and 2016 showed that MAXSER outperformed all other approaches before transaction costs and all except the naive $1/N$ and one version of the global minimum variance approach after transaction costs. Empirical analysis with the S&P 500 Index constituents over the same period revealed that MAXSER outperformed all other approaches before and after transaction fees. Appendix D.2 covers the implementation details of the MAXSER approach.

6.2.2 Robust Portfolio Choice

The 2021 paper *Robust Portfolio Choice* by Valentina Raponi, Raman Uppal, and Paolo Zaffaroni introduced the robust-mean-variance ("RMV") approach that was based on the creative idea of constructing two inefficient "alpha" and "beta" portfolios that collectively produced an efficient mean-variance portfolio.[21] Consistent with intuition, the beta portfolio depended on factor risk premia and the alpha portfolio depended on pricing errors.

Raponi, Uppal, and Zaffaroni performed simulation analysis with parameters for generating returns calibrated from the Dow Jones 30 Index constituents and empirical analysis using monthly returns of the constituents of the Dow Jones 30 Index and randomly selected 100 constituents of the S&P 500 Index.

They compared the RMV approach to the mean-variance, global minimum variance, the $1/N$, and the MAXSER approach discussed in section 6.2.1. The simulation analysis with 30 assets showed that the RMV approach delivered a Sharpe ratio of 0.932, which is 123 percent higher than that of the $1/N$ approach and 12 percent higher than that of the MAXSER approach. When the number of assets in the simulation was increased to 100, the RMV approached outperformed the $1/N$ portfolio by 175 percent and the MAXSER approach by 30 percent.

Empirical analysis with the Dow Jones 30 Index constituents for the period between 1977 and 2016 produced an out-of-sample Sharpe ratio of 0.872 for the RMV approach outperforming the $1/N$ approach by 161 percent and the MAXSER approach by 105 percent. Empirical analysis with S&P 500 returns revealed similar relative outperformance for the RMV approach. It produced a 1.222 Sharpe ratio, which was both economically greater than the 0.494 Sharpe ratio of the $1/N$ approach and the 0.672 Sharpe ratio of the MAXSER approach and the differences were statistically significant.

The authors argued that the RMV approach was superior to the $1/N$ approach because it combined the alpha and the beta portfolio, whereas the $1/N$ approach was a proxy for the beta portfolio. The RMV approach was also superior to the MAXSER method because the latter approach was designed to have mostly zero alpha portfolio weights and a small number of relatively small non-zero alpha portfolio weights whereas the former method was designed to take full advantage of the alpha portfolio. Appendix D.3 briefly describes the key ideas and the implementation steps of the robust-mean-variance approach.

6.3 INTERESTING NUGGETS

This section shares several interesting nuggets: a Bayesian risk parity approach, the *empirical Bayesian approach* of Michaud, performance evaluation with funding ratios, and investing in a low-yield environment.

6.3.1 Bayesian Risk Parity

Section 5.1.3.2 introduces the Black-Litterman approach that uses the following process:

1. Capitalization-weighted market portfolio is a good starting portfolio for any investor according to the Capital Asset Pricing Model.
2. Reverse optimization produces the implied expectations of assets' performance.
3. Expert opinions are defined in terms of expectations about absolute or relative performance of assets with a certain degree of confidence/uncertainty.
4. Black-Litterman framework incorporates the expert opinions and produces a new set of portfolio weights.

The 2017 study *Black-Litterman, Exotic Beta and Varying Efficient Portfolios: An Integrated Approach* by Ricky Cooper and Marat Molyboga highlighted two weaknesses of the classic Black-Litterman approach related to the steps 1 and 3 and offered potential solutions:[22]

- First, while the market portfolio generally performed well and was rooted in economic theory, empirical research showed that it was not optimal. For example, in his 1977 paper *A Critique of the Asset Pricing Theory's Tests Part I: On Past and Potential Testability of the Theory* Richard Roll argued that the true market portfolio was unobservable.[23] The aforementioned 2009 study *Optimal versus Naive Diversification: How Inefficient is the 1/N Portfolio Strategy?* showed that the naive 1/N outperformed the capitalization-weighted portfolio.[24] Finally, the 2012 paper *Leverage Aversion and Risk Parity* showed that the capitalization-weighted market portfolio was not efficient and argued that the risk-parity was more efficient due to leverage aversion.[25] Moreover, capitalization-weighting is not feasible for hedge funds since AUM is a poor proxy of capitalization. Thus, an equal-risk portfolio is an excellent candidate for a starting portfolio.

- Second, expert opinions could be difficult to acquire or costly. Thus, Cooper and Molyboga proposed using exotic betas such as low-volatility anomaly or momentum as opinions.

This methodology can be applied to hedge fund portfolios. A risk parity approach can be used as a starting portfolio as suggested by Cooper and Molyboga. However, other options may include $1/N$ or the *HRP* approach of Marcos Lopez de Prado.

We consider two popular opinions regarding hedge fund performance:

- **Equal Sharpe ratio**: All hedge funds perform similarly and their true Sharpe ratios are equal.
- **Risk-adjusted momentum**: There is performance persistence among hedge funds and past winners have higher expected Sharpe ratios than losers.

Appendix D.4 provides a detailed technical description of how the framework can use the risk parity portfolio as the starting portfolio and incorporate the equal Sharpe ratio or risk-adjusted momentum opinions.

6.3.2 Empirical Bayesian Approach of Michaud

The 1989 study *The Markowitz Optimization Enigma: Is Optimized Optimal?* argued that mean-variance optimization was an "estimation-error maximizer" and produced unintuitive portfolios because they didn't make investment sense and failed to provide investment value.[26] Bayesian techniques such as shrinkage or Black-Litterman method discussed in Section 5.1.3 explicitly account for estimation error, but they depend upon assumptions ("prior") that could be inconsistent with the empirical data. For example, as discussed in the aforementioned 2017 study *Black-Litterman, Exotic Beta and Varying Efficient Portfolios: An Integrated Approach* the Black-Litterman approach relied on the market portfolio as the starting point ("prior") but empirical studies showed that risk-parity was a more efficient portfolio.[27]

In his 2005 study *Bayesians, Frequentists, and Scientists* Bradley Efron suggested an "empirical Bayes" approach that relied on data to derive the important assumptions ("prior") and then used those data-driven assumptions within a Bayesian framework to mitigate estimation error.[28] The resampling methodology of Michaud that follows the empirical Bayes principles was described in detail in the 2008 paper *Estimation Error and Portfolio Optimization: A Resampling Solution.*[29] They key idea is simple. Classic mean-variance approach creates a portfolio that is optimal for a single historical path, which is unlikely to repeat, leading to poor out-of-sample performance discussed in Section 5.1.2. By contrast, Michaud's method resamples historical data to create a large number of potential paths (alternative universes) and produces a portfolio that performs well across the scenarios. Specifically, the framework creates an efficient frontier for each path and blends them together to create a resampled efficient frontier.

Several studies showed that resampled, or empirical Bayes, portfolios were superior to the Bayesian solutions. For example, in their 2003 study *Resampled Frontiers versus Diffuse Bayes: An Experiment* Harry Markowitz and Nilufer Usmen used simulated data to show that resampled portfolios outperformed the Bayes portfolios with diffuse priors, on average.[30] The 2008 study *Bayes vs. Resampling: A Rematch* by Campbell Harvey, John Liechty, and Merrill Liechty refined the analysis by Markowitz and Usmen by employing the Markov Chain Monte Carlo (MCMC) algorithm recommended in the Bayesian literature. They found that the MCMC algorithm improved the performance of the Bayesian approach, yet the Bayesian solution outperforms the resampled approach only when future distribution of asset returns closely resembles their historical distributions.[31] The just cited 2008 study *Estimation Error and Portfolio Optimization: A Resampling Solution* pointed out that Bayesian estimation of inputs and resampling were complementary techniques that could be combined into a single approach.

Since hedge fund returns typically follow a non-normal distribution and often exhibit positive serial and cross-correlation, we suggest a few modifications of the original resampling approach:

1. Standard deviation may not be the most relevant measure of risk. For example, conditional value-at-risk discussed in Section C.5.2 or conditional expected drawdown discussed in Section C.5.3 can be considered as alternative measures of risk for efficient frontiers.

2. Data sets that are used to estimate efficient frontiers should preserve the important characteristics of hedge fund returns whether relying on bootstrapping or Monte Carlo simulations:

   - Block bootstrapping discussed in Appendix B.1 is particularly relevant for hedge fund portfolios because it preserves skewness and kurtosis in returns as well as the serial and cross-correlation characteristics of the original data set.
   - Monte Carlo simulation should include higher moments such as skewness and kurtosis to capture non-normality in returns and retain the serial and cross-correlation characteristics.

6.3.3 Portfolio Contribution with Funding Ratios

As the stocks and bonds were in a bear market at the end of 2022, the funding ratios of public pension fund portfolios have been plummeting. "The overall estimated funding ratio of the 100 largest U.S. public pension plans fell to 75% in August due primarily to negative investment returns for the month, according to the Milliman 100 Public Pension Funding index."[32] Although funding ratio is the most direct and relevant measure of the pension fund's ability to meet its obligations, most studies rely on other performance measures. The 2019 article *Commentary: Evaluation of Alternative Investments in Pension Fund Portfolios* by Marat Molyboga closed that gap by proposing a simple intuitive methodology designed to evaluate the contribution of any investment to a pension fund's portfolio using funding ratios.[33]

The author illustrated the methodology by considering an investment decision of allocating to CTAs. The Societe Generale Trend Index, an index comprised of 10 largest trend-following CTAs open to new investments, was used to represent the CTA investment.

Molyboga found that a modest 10 percent allocation to CTAs consistently improved the funding ratio both during times of stress

(2001–2002 and 2007–2008) and during normal times by 8 percentage points, on average. The proposed methodology can be applied to evaluate hedge fund allocation decisions.

6.3.4 Investing in a Low-yield Environment

Although hedge funds are typically evaluated relative to a factor model, investors tend to make global investment decisions with a framework of strategic asset allocation that relies on forward-looking return expectations for individual factors. Given the low-yield environment of 2010–2021 with the stock market at all-time highs and bond yields close to all-time lows, institutional investors wanted to know how to reposition their portfolios including hedge fund portfolios for success in the low-yield environment.

The opinion piece *Hedge Funds: Coping with Low Interest Rates* by Michael Going and Marat Molyboga published in Investments & Pensions Europe introduced a framework that provided practical guidance to institutional investors for making strategic asset allocation decisions that could succeed in a low interest rate environment.[34]

The authors contributed to the debate in three ways:

1. Defined three market scenarios that are especially relevant for investors going forward: spiking interest rates that typically last between six months and two years; gradually rising interest rates that tend to occur between two and five years; and coinciding periods of rising rates and falling equities.

2. Identified historical sub-periods between January 1961 and December 2020 that matched the scenarios.

3. Evaluated the performance of assets and strategies during those historical periods. The framework was illustrated by considering U.S. stocks, U.S. government and corporate bonds, and commodities, to represent the bulk of asset classes used by institutional investors. Going and Molyboga also examined the five risk premia of time-series momentum, value, cross-sectional momentum, carry, and defensive often employed by quantitative hedge funds. Using risk premia performance rather than the track records of individual hedge funds or hedge fund indices solves the problem of short track records and the diversity of hedge fund strategies.

The analysis was performed using excess returns for the assets and risk premia available in the AQR data library. However, the framework can be expanded and customized by including additional asset classes, risk premia or any other investments relevant for specific institutional investors.

The authors shared the following interesting findings:

- Investors should be concerned about the performance of their stock and bond portfolios in a low interest rate environment. In particular, gradually increasing interest rates or a combination of rising rates and falling equities pose major performance threats to these types of portfolios.

- Although commodities have been heavily criticized in recent years because of their poor performance, they may provide rare value in a low-yield environment. Commodities are known for their inflation hedging characteristics, and they performed extremely well during the most challenging periods considered in the study.

- It is prudent to consider diversifying strategies such as momentum, value and carry that are commonly employed by quantitative hedge funds because they have consistently delivered positive performance across all the market scenarios under examination. However, the defensive strategy tends to struggle in a low interest rate environment.

- Finally, time-series momentum stands out as another potential offset against portfolio losses within this environment. While our study shows that it performs well in a low rate environment, it is also known for producing superior returns during market crisis (crisis alpha) as evidenced during the global financial crisis of 2008.

Although the low-yield environment ended in 2022, inflation is at the highest level in 40 years and the essential lesson of coming up with forward-looking expectations by analyzing historical periods of similar environment rather than relying on the most recent 10, 20, or 40 years remains. As Carlos Slim Helu said: "With a good perspective on history, we can have a better understanding of the past and present, and thus a clear vision of the future."

6.4 KEY TAKEAWAYS

Following are the key takeaways from this chapter:

- Machine learning approaches such as the HRP and the MHRP are consistent with the top-down perspective of institutional investors who build their portfolios by starting at the asset class level and then going down to the level of the individual securities. The topic of cleaning correlation matrices is an excellent area for future theoretical research with practical implications.
- MAXSER and robust portfolio choice are two cutting-edge methods that should be considered by hedge fund managers and investors.
- Other interesting approaches include Bayesian risk parity and Michaud's resampling.
- Pension plans can benefit from explicitly evaluating the impact of candidate investments on their funding ratios.
- When market environment changes, historical periods with a similar environment rather than the most recent period are more relevant for forward-looking portfolio decisions.

CHAPTER 7

Expert Hedge Fund Managers

"Nature is written in mathematical language."—Galileo Galilei.[a]

"What counts in life is not the mere fact that we have lived. It is what difference we have made to the lives of others that will determine the significance of the life we lead."—Nelson Mandela.[b]

The first six chapters of the book presented the empirical evidence from hundreds of research studies published in peer-reviewed academic financial and mathematical journals. We continue our journey by turning from academic papers to exceptional individuals. We believe that you will benefit from their valuable insights that are based on decades of quantitative research and reflection. This chapter relates personal stories of expert hedge fund managers, including positive and negative experiences, past and present challenges, and many opportunities for all of us to learn and be inspired. This chapter also provides a deep dive into four hedge fund strategies: trend following, machine learning, emerging markets, and sustainable investing. For those who are interested in other types of hedge fund strategies, we recommend the book *Efficiently Inefficient: How Smart Money Invests and Market Prices Are Determined* by Lasse Pedersen.[1]

[a]Galileo Galilei was an Italian astronomer, physicist, and engineer famous for advocating that the Earth was orbiting around the Sun.

[b]Nelson Mandela is regarded as a symbol of democracy and social justice, a Nobel Peace Laureate, 1993.

DOI: 10.1201/9781003175209-7

7.1 TREND FOLLOWING WITH KATY KAMINSKI

As Chief Research Strategist at AlphaSimplex, Dr. Kaminski conducts applied research, leads strategic research initiatives, focuses on portfolio construction and risk management, and engages in product development. Dr. Kaminski is a member of the investment committee. She also serves as a co-portfolio manager for the AlphaSimplex Managed Futures Strategy. Dr. Kaminski joined AlphaSimplex in 2018 after being a visiting scientist at the MIT Laboratory for Financial Engineering. Prior to this, she held portfolio management positions as a director, investment strategies at Campbell and Company and as a senior investment analyst at RPM, a CTA fund of funds. Dr. Kaminski co-authored the 2014 book *Trend Following with Managed Futures: The Search for Crisis Alpha.* Her research and industry commentary have been published in a wide range of industry publications as well as academic journals. She is a contributory author for both the CAIA and CFA reading materials. Dr. Kaminski has taught at the MIT Sloan School of Management, the Stockholm School of Economics and the Swedish Royal Institute of Technology, KTH. She earned a B.S. in Electrical Engineering and Ph.D. in Operations Research from MIT where her doctoral research focused on stochastic processes, stopping rules, and investment heuristics.

Katy, you earned your Ph.D. in operations research from MIT, which has one of the best programs in the world, under Andrew Lo, who is one of the greatest minds in finance. That sounds very challenging. Could you please tell us about your experience at MIT?

MIT is awesome. It was a wonderful experience. I couldn't get enough of it. I was there for 10 years. I loved MIT because I had to use math a lot and I had to learn all the time.

I started majoring in electrical engineering during undergrad. I got interested in finance after an internship at a French bank. The internship was very fast paced, and I really enjoyed doing modeling. I started thinking that finance could be a lot of fun, particularly if I could continue using math.

I decided to do a Ph.D. in Operations Research because it's about using Math techniques to solve challenging problems, which is awesome for a girl who loves math. When I met Andrew Lo, I got very lucky. He needed a teaching assistant and I needed an advisor. He was also doing research on rules-based investing. My Ph.D. thesis was focused on trying to understand the systematic rules that investors use, which was more operations research rather than a finance question at the time.

What is very interesting about Andrew Lo's research is that he often focuses on the intersection of finance and other fields whether bringing quantitative approaches to trading or bringing quantitative approaches and finance into healthcare, pharmaceuticals, and biotech. That's where interesting things will happen in the future. Innovation is always at the intersection of different fields.

Andrew Lo is brilliant, but he is also very kind. I appreciate him as a mentor because of his mentality of finding interesting people regardless of their background or personality and collaborating with them to contribute and come up with something new and innovative. People often say that they aspire to do that, but he lives it.

It sounds as if you decided to go into finance because it was an interesting area to apply math.

Yes. I started off wanting to do finance because I thought it was a great place to do more math. However, my mom also had background in finance. She worked in financial planning and often talked about the stock market and earnings per share. I was not intimidated by finance because my mother was good at finance. She was the one in my family watching CNBC and Bloomberg all the time. I didn't get it until I started doing the internship. That's the link that ignited the fire for me to be interested in finance. First, it was an interesting place to try to model uncertainty and things that are complicated. Then after I got a taste of it, I started thinking that it was also very practical and relevant for everyday life. I have been more interested in finance every year since then.

You have taught at MIT, Stockholm School of Economics, and the Swedish Royal Institute of Technology. What do you enjoy about teaching?

I love teaching because it's an opportunity to connect with younger people. It is also a way for me to give back to others.

I have also always believed that if you can teach something to someone else, you can understand it. I think it's an important skill. In finance, the truly helpful people are those who can make things easy rather than complex. It is easy to be complicated, it is hard to be simple. There is a great skill in being able to distill complex things down into very understandable units. Teaching is about practicing doing that over and over.

Another reason why I teach is because I consider it to be my duty as someone from an underrepresented group in our space. It's important for people to see someone they can relate to in positions that they aspire to. For example, teaching in the master's in finance program at MIT as a female portfolio manager is a great message for all students, not just the female students. I have an interesting job, I work on interesting projects, and I have something interesting to share.

I want to continue teaching because it is great to continue developing the skill and because it's a great way to reconnect with the next generation and give back to people.

Do you find sometimes that students ask interesting questions that spark ideas for your own research?

It does a little bit in projects. I have done a lot of project sponsorship for the University of Massachusetts through Mila Getmansky. She teaches courses on alternatives and hedge funds. Each project has a team of students who try to solve a challenging problem such as investing in an inflationary environment or creating a new futures market. It is a lot of fun when you can share a new idea, and watch a group of students start from scratch to tackle it. I really enjoy being part of the students' journey into finance.

You have been active in academia and the industry for many years. How would you compare them? What is the gap between them? What can practitioners learn from academia? What can academics learn from the industry?

I have worked a lot in this space, particularly in Sweden. Industry and academia have always had a little bit of a divide, but it's been fabulous to see over the last 20 years the divide narrowing. I am seeing industry incorporating ideas from academia more proactively. For example, AQR has done a lot of work to help bridge the gap. I am seeing a lot of firms recognizing the value of an academic approach incorporated in what they do to help educate and structure thinking, define new strategies and benchmarks, and even provide a new nomenclature. We have seen a lot of that with alternative risk premia.

I think the gap always exists based on the fact that they have different objective functions. Academics try to get published in a short list of journals. The prowess within that group is based on the number of publications in top journals, and people who get published there are also on the editorial committees of those journals.

Mila Getmansky wrote an interesting paper in the Journal of Finance about the importance of networks and female participation in academia. She showed that the network effect is a major challenge for smaller groups and minorities. That's very true within the finance world of academia. It is a very well-connected network. The connections and the feedback of the connections are very important for the cycle of publishing.

There is a divide because we have a different objective. Our objective is the fiduciary duty. It's not the most exciting problem that will get published. It's about providing the best service to our clients and find the best investment solution for the objective of our clients. It's a very different dynamic than academia. I like industry more, but I have tremendous respect for academia and the amount of work and effort that goes into that endeavor.

Although the objective function is different, it sounds as if practitioners can still learn from academia about the way academics approach solving problems and the way they impose frameworks. If someone understands academic work, they can be more successful at implementing solutions for clients. Do you agree with that?

You also need to understand the academic work to understand its pitfalls. That's what I think is the most fascinating. I cannot tell you how many times I have received an email from a salesperson who sends me a paper and highlights the fact that it is an academic paper. I always tell them that we have to think about why the paper is referenced in an industry setting. There is always an underlying objective and a reason for why the paper is shared.

In the industry, we have to have ideas that walk forward. In academia, people look at ideas that have worked in the past. That is a very different world. Those two types of ideas are not always compatible. Some spurious relationships are fun to write about in a paper, but they are unlikely to repeat themselves in the future. I appreciate having a Ph.D. because it provides a certain level of maturity, sophistication, and humbleness that helps in finance because the real world is so dynamic and there is so much noise. It quickly teaches a lot of humility.

I have always liked the trend following strategy because it is so simple that it should not work from an academic perspective but from a behavioral and a parameter robustness perspective it makes sense.

You have been very successful in the industry, but your journey was probably not always easy. Can you please share some of the obstacles that you had to overcome along the way?

I have always done what I really liked, and I have had plenty of moments when people did not believe in me. I have had plenty of moments when I was underestimated. One of my great appreciations is that I have always tried to take underestimation as a power rather than a thorn. I have just acknowledged that when people underestimated me that it gave me an extra power if I was willing to overcome that. It is not easy. There have been times when people believed that I did not know math even though I had a Ph.D. in operations research. I believe that in every weakness there is strength. The number one strength is to acknowledge how to use that as a force to make yourself stronger instead of being disappointed.

For example, if you feel like you are getting underestimated for some reason, you either have to change your expectations or you have to learn to look at things in a different way. It's important to surround yourself with the right people, find the right allies, and not try to convince people who are not convincible. It is also important to continue to be passionate about what you like to find a pathway to success.

I am working on the book *From Exception to Exceptional.* We all feel different whether underestimated, undervalued, left out, or marginalized. The key question is how turn that into something positive, which is a really important challenge for people. It is difficult to move forward and not be disappointed. When something is frustrating, it is still possible to have a positive perspective. For me it's always about having a good attitude, be positive, but also surround myself with real people who support me.

Being underestimated is a superpower because you have an opportunity to excel. If you are underestimated, it is easier to exceed the benchmark of expectations. The problem is that we want that to happen instantaneously. In life it's important to win your battles when you can. You can always look for allies and build a support network of people who are on your team. As long as you have a team, you are not alone.

When I was in undergrad, I went to Polytechnique, a school with 5 percent female at the time. I was studying physics. I lived in a men's dorm because they did not have room in the one hall they had for women. I was all by myself but I don't remember feeling totally isolated. I got used to it, I found allies. Once you get the right perspective, you can overcome anything.

It is so important to give back and support young people because you can become part of their support network. That's really what they want and need.

I also had an interesting experience with my friend Mila Getmansky. We were at a conference in New York, and there was a paper discussed about performance attribution bias. This paper showed that in economics women who wrote papers with men were less likely to get tenure than their male co-authors. It was a controlled study. Women who wrote papers alone got full credit for their work. I was shocked. I realized that I had written most of my papers alone. Mila and I were laughing because we got lucky. Our strategy of writing papers alone was the right strategy. Sometimes we think that if we write a paper with a senior person who has a lot of respect, it might help us gain some respect too. The paper was showing that that was not the case. I don't like biases, but you have to be aware of them.

Then when I wrote my book about trend following with Alex Greyserman, I cannot tell you how many times I met people who assumed that I didn't write it. I just laugh because I know the bias. Once again, instead of getting angry at them, I laugh about the attribution bias.

Your book Trend Following with Managed Futures: The Search for Crisis Alpha *is the bible of trend following. It presents a lot of empirical evidence with practical implications for hedge fund managers and investors. Why did you decide to write this book?*

I was fascinated by the topic, and I had written a lot of articles about trend following. There is no better way to become an expert on something than to challenge yourself to do something hard like writing a book.

The whole story is funny. When I was in the states visiting Andrew Lo and MIT, I had a meeting with Alex Greyserman in New York. I knew him because I had read his papers and he had read mine. We were naturally very similar in the way we thought about things. We went out for lunch. It was our first meeting. It was so nerdy. We had so much in common and so much respect for each other. I looked at him and said: "We should write a book." We decided to do it.

We lived on two different continents. I lived in Sweden, he lived in New York. I came back with an outline. He got connected with someone from Wiley, the publisher. Just like that we were on the hook to write the book. We talked on the phone every day. We met in London and Stockholm when we could. We finished the book in a year.

It was a great experience. I learned so much from the book. I loved the challenge of answering questions that I didn't know the answer to. It's very hard! It's easy to have a plan. It's hard to have a destination and not know how to get there.

When you look at your book, what are you most proud of?

Finance is a field where terms are well known. I remember feeling overwhelmed that I came up with crisis alpha, which is a new term in finance. Regardless of whether people like it or not, it seemed crazy for a young person in the industry to come up with a new term. The Chicago Mercantile Exchange group had the highest number of downloads of this paper out of all papers on their website. They were very excited about it. I got a chance to meet a lot of interesting people, which eventually led to writing this book.

I was excited about this topic because I was bothered by the dichotomy between academia and the industry. At that point trend following was considered a voodoo strategy. People thought that trend following was dumb, and it didn't work. I was upset to hear that because trend following had worked for decades, and I wanted to show the reason behind it.

The book talked not only about why trend following strategies worked, but also how institutional investors could get comfortable and benefit from those strategies. Our book was different from the previous books on trend following because our book was amenable to institutions.

Your crisis term is used so widely in the financial industry. How do you define it?

The original definition of crisis alpha was the potential profit opportunity during periods of market stress or crisis. It means finding opportunities during distress and any strategy that can maneuver itself to benefit from such an environment. This term is not specific to trend following. Trend following is one of the strategies that tends to have some of those attributes, but not necessarily all of them.

Your career has primarily been in the trend following space that is often described with your concept of crisis alpha. Why are you passionate about trend following? Why should hedge fund managers and investors consider trend following?

Trend following is about applying mathematical techniques to measure where markets are going and what the markets are doing, and not necessarily why they should go somewhere or what they should do. Trend following strategies are about human behavior. As a strategy it will always have a place because there is so much uncertainty in finance. We never really know where the world is going. Sometimes we are wrong, but there is information in the behavior of people. That has been particularly interesting over the last three years because we have seen the world change, but the information incorporates much slower than what we would expect if markets were perfectly efficient.

It is always exciting because there is always a trend somewhere. It is not a high Sharpe ratio strategy, but it is robust. Trend following has stood the test of time, and it can be very oppositional to the traditional portfolio positioning from time to time providing something that is very different. For example, shorting bonds is something that is very difficult for people to do without the rules of trend following.

I know that you are passionate about diversity and you mentor women who want to succeed in the hedge fund industry. What drives that passion and what are some of the key areas that you tend to focus on as a mentor?

Our problem is that we often try to put a square peg in a round hole and a round peg in a square hole. We try to use the same strategies that have worked with the same problem and don't solve the problem. That's one of the issues with diversity. From my experience, women and men sometimes react differently to risk environments, but our interview process and culture are not necessarily productive at driving leadership in the underrepresented groups.

We have to acknowledge the power of the size and the demographics of a team. The incentives are not necessarily aligned with mutual support. Our space is very competitive. The interview process is very aggressive and focused on solving problems. If you don't acknowledge some of the biases in the way that women present themselves, you may miss the mark. For example, women tend to be very honest about their true competencies. If they know four out of five programming languages, they would be honest about not knowing the fifth one. Male candidates tend to say that they can do anything. The problem with that from an interviewing perspective is that those two answers are typically interpreted very differently. It's important to be aware of those biases and themes.

Women are less overconfident. When they look for a job that has six different criteria, they will expect themselves to meet all six of them. They want to check all the boxes. A distribution for men is wider. When employers look for someone, they tend to look for a prince charming. We want to find someone who is fluent in SQL, has a Ph.D., has such and such attributes. For employers, it's a laundry list of attributes, but job candidates, particularly women, don't think about the criteria the same way. If you want to promote diversity, you have to look at all parts of the process including the job criteria. For example, you can't say a Master or Ph.D. degree because you will get Master and Ph.D. applicants who are male, but you will not get any Master applicants who are female. When I tell women about that, they laugh because they know that they do it.

It has been very interesting for me now because a third of my team is female. The dynamics are very different. It has been very interesting to work with junior women on the team. I have found that it is helpful to give them some time to get comfortable with the environment, encourage them, and to show them that it is ok to not know everything. I would rather get an honest humble answer, but that's not necessarily what gets rewarded during interviews.

When I mentor women, I tend to focus on some of those tendencies and biases that we may not acknowledge. For example, women are less likely to talk to recruiters. They tend to be more concerned about risk in changing positions. When I found that out early in my career, I had to override it and make myself talk to recruiters. I tell junior women that they lose information if they don't talk to recruiters. They lose competitive information about where they are in their careers but also opportunities and trends. That is a disadvantage. The reason why women don't talk to recruiters is because they wait until they need to change jobs to get information and they may feel guilty to have those conversations. I try to overcome this bias by always having those conversations and providing something valuable to those individuals.

Even if I am not looking because I love my job, I can have a conversation with a recruiter and introduce him to five other people.

There are many themes like that. If you acknowledge them, you can overcome them and use them to your advantage. If you ignore them completely, you may not be aware of how they impact you. That is a big issue in any situation where you are an exception to the rule.

As you know, this book is about quantitative hedge fund investing. What advice would you give to hedge fund managers and investors that can help them improve their performance?

The world is changing. The most important thing going forward is different from what was important in the past. Our space is extremely diverse. I see that when I talk to junior people because they come to me and tell me that they want to do what I do. What I do is a very small piece of the huge financial puzzle. There are so many things that people can do. There is such a wide range of different hedge fund strategies and interesting investment areas.

People succeed in our space when they find a place where they have a unique edge, or they use a different methodology or an approach in an area that is uncommon. It's about finding that niche. It's important to find a place where you find passion, but also where you can use new techniques and outside perspectives to improve or create something new.

7.2 MACHINE LEARNING WITH KAI WU

Kai Wu is the founder and Chief Investment Officer of Sparkline Capital, an investment management firm applying state-of-the-art machine learning and computing to uncover alpha in large, unstructured data sets. Prior to Sparkline, Kai co-founded and co-managed Kaleidoscope Capital, a quantitative hedge fund in Boston. With one other partner, he grew Kaleidoscope to $350 million in assets from institutional investors. Kai jointly managed all aspects of the company, including technology, investments, operations, trading, investor relations, and recruiting. Previously, Kai worked at GMO, where he was a member of Jeremy Grantham's $40 billion asset allocation team. He also worked closely with the firm's equity and macro investment teams in Boston, San Francisco, London, and Sydney. Kai graduated from Harvard College Magna Cum Lauda and Phi Beta Kappa.

Why did you decide to go into finance?

I am a son of immigrants. My mom is an artist, and my dad is a doctor. I grew up without knowing anything about finance. I had never picked up the Wall Street Journal. I had never seen a movie about Wall Street. I didn't know anything about Warren Buffet.

I went to Harvard thinking that I would major in political science. While at Harvard, I realized that Economics was a just a quantitative version of social sciences, which was attractive to me. I started studying economics, and then the Global Financial Crisis happened. All the banks started failing, and that got me interested in what was potentially driving all the calamity.

I ended up doing my senior thesis with Ken Rogoff (a well-known Harvard Professor who served as Chief Economist and Director of Research at the International Monetary Fund in 2001–2003). Ken specialized in financial crises. The overall idea of my thesis was that most of the time markets are not in equilibrium. Imbalances in the system—whether external, fiscal, monetary, asset price, or credit—build up and then have to unwind, causing crises. We trained an econometric model on hundreds of years of data and dozens of financial crises and found that these imbalances tend to precede financial crises. This project got me interested in the financial markets.

Around that time, I also got my first internship at Grantham, Mayo, and van Otterloo (GMO). When they called me for an interview, I had to google to learn about what they did, because I didn't know much about the investment industry. There I met Edward Chancellor, who wrote the book *Devil Take the Hindmost: A History of Financial Speculation.* He became an unofficial mentor on my thesis. After graduating from Harvard, I decided to join GMO full-time.

I worked with Ed on a lot of interesting macro projects. Then I joined the asset allocation team. I had a really good experience there. That's how I got into finance. It was not deliberate. It was accidental. I guess the better question is why I decided to stay in Finance. I love my job. I wake up every morning loving what I do. There are always new challenges.

At some point, I left GMO and went on an entrepreneurial path, which led to additional fun challenges. Markets are so dynamic and competitive, intellectually interesting. There are so many ways to play the game, whether you are a discretionary trader, fundamental analyst, or a quant. Of course, there is also great purpose in investing. Whether at GMO or now, I am driven by the desire to perform for my clients.

There is more purpose in being an investor that is often portrayed in the media.

Can you please share some of the obstacles that you had to overcome along the way?

Nothing is easy, but I am sure there are many people out there who had much harder roads than me. My relative lack of experience and knowledge of the investment industry was a hindrance in entering the field. However, I would say that the obstacles became way more significant after I left GMO and embarked down entrepreneurial road.

I joined a former GMO colleague and cofounded a new hedge fund in Boston. The hedge fund grew to a few hundred million dollars in assets under management. Then after I left the hedge fund, I started my own firm Sparkline. That process of entrepreneurship launching two firms and multiple funds is very different from being in a large organization. I don't want to glorify entrepreneurship because it is extremely hard and very stressful. When I left GMO, I was a good researcher, but I had no experience in fund raising, operations, trading, and management. These areas were completely new to me. This forced me to step out of my comfort zone.

For example, in fund raising, it is typical to get rejected in 99 out of 100 meetings. That's a lot of rejection, but you have to get up in the morning and keep working. Some of the best advice I got about entrepreneurship was from my former boss, Jeremy Grantham. He told me: "One of the best things you can do when you are young is to build a war chest." It is important to save money. To this day Jeremy is very frugal, in a good way. When I left GMO, I had some money saved up. By the time my first firm got profitable, I had zero dollars in the bank!

That's the major lesson of entrepreneurship. So much is out of your control. You don't have a timeline or a roadmap to when you are going to be profitable. You need staying power, which includes both financial and mental resources. It takes a lot of patience and willingness to grind it out and wait because things can turn around any moment. You could get lucky and be profitable on the first day, but it could just as likely take a year or 10 years. So the major piece of advice I would give to someone interested in this path is to have staying power.

Today machine learning seems to be everywhere. Self-driving cars, medical diagnosis, fraud prevention. However, it's still not well understood. What is machine learning and why are you so passionate about it? Why should hedge fund managers and investors consider machine learning?

Machine learning is central to quantitative finance. If you think about it—*linear regression*, the cornerstone of empirical finance, is a form of machine learning. Once you realize that, you recognize that you already are using machine learning. Don't be afraid of machine learning!

Then, the question becomes how much complexity you want to introduce to the model. Non-linearities and interactions can be incorporated by adding terms to your regression. Or you can use *decision trees*, *random forests*, *boosting*. Or you can go all the way to extremely complex deep learning models. There is a tradeoff as you move from linear regression, which is the simplest approach, to more and more complex models. While complex models can capture more nuances in the data, they require a lot more data to train. If you try to train a model with 1,000,000 degrees of freedom on 1,000 data points you will almost certainly overfit.

Ultimately, you have to pick the right tool for the job. If you are trying to drive a nail into a table, you don't have to bring a sledgehammer. A lot of people whom I have met in the industry use cool tools for the sake of using cool tools. I think this is a mistake. However, that does also not mean that you should only stick to only using the old tools, which incorrectly assume that all relationships are linear. You don't want to ignore the richness of this world.

I like categorizing data as structured and unstructured. Structured data is the type of data that you would find in Excel spreadsheets or SQL databases. For example, time-series of returns, trading volume, and price-to-earnings ratios. Unstructured data is different. For example, text, images, audio, or videos. There is a huge divide between those two realms.

I wrote a paper about deep learning that shows the difference. When you work with structured financial data, you should be very careful about *overfitting* because it usually doesn't have many degrees of freedom. For example, macroeconomic data is quarterly and exhibits a high degree of autocorrelation. What's the real breadth you have there? Financial data is super noisy. The true dimensionality is lower than you might think. As a result, as you increase complexity, you should be very careful to not overfit. I showed a simple example in the paper where I looked at one, two, three, four, and five-layer networks. A one-layer network is just a linear regression, and then it gets more and more complex with each layer. What is the optimal complexity? It was three for the dataset. It's important to recognize that a three-layer model is millions of times less complex than the state-of-the-art models for text processing. You tap out very quickly. My general advice when working with structured data would be to stay with parsimonious boosting and tree-based models that are more robust to overfitting.

However, when you work with unstructured data, it is a totally different game. You have access so much data. I can download the whole English Wikipedia and use it to train my model. It is an enormous amount of data. Moreover, you don't have to worry as much about non-stationarity. The English language evolves only very slowly over time. As a result, you can get very good results with more complex models. Google switched from a complex rule-based model to a deep learning model called BERT. This model produced better results, while requiring much less code and manual training. It is incredible that you can replace hundreds of thousands of man-hours of human linguists with a simple statistical model. When it comes to *natural language processing* (NLP) or computer vision, deep learning is here to stay.

I think there is a lot of value of taking unstructured data, bringing it into the structured realm using deep learning models, and using it within a standard quantitative framework alongside traditional market data.

What are some examples of unstructured data that can be used in finance?

Primarily text data. Certainly, there are examples of people using vocal stress and images of CEOs to analyze their body language. That's out there, and maybe that works. However, 99 percent of what I have seen relies on text data.

If you are a fundamental analyst working for a discretionary hedge fund and you are analyzing retail, you are probably reading 10-Ks of the target company and its competitors. You are probably going to read through the business description and the risk factors to understand the business better. You will probably cover the Management Discussion and Analysis (MD&A) section to learn about their outlook for the business. You probably go through the last few quarters of earnings calls to learn about what they have been talking about. You can do those things with a machine. While machines often have a less deep level of understanding, they are very adept at connecting the dots across thousands of companies in a way that a single human can't.

Transfer learning is an interesting topic in machine learning. The more complex the model, the more data it needs for training because it has a very large number of parameters. If you want to train a model with a billion parameters on 10-Ks, it might be very challenging because you may only have three thousand 10-Ks. You can overcome that issue by first training the model on Wikipedia, the general-purpose corpus. You will find that the model starts converging, but it's not quite there yet. The model knows English, but it doesn't know the financial jargon. You can take that model and fine tune it on your 10-Ks to take it across the finish line. Now your model doesn't only know English, but it also knows that EBITDA means "earnings before interest, taxes, depreciation, and amortization"!

I find that most hedge fund managers claim to use some version of machine learning. What advice would you give to investors who want to separate experts from pretenders?

Well, this is true if you count linear regression. Of course, there is a wide variance in sophistication in machine learning. So what questions should you ask? First, I would suggest there are a few questions that you should not ask. I don't think you should care how many Ph.Ds. or how many years of experience they have. It doesn't matter if somebody has 50 years of experience, because most of this technology was developed in the last few years and wasn't taught in formal programs until recently.

With that said, there are helpful questions to ask. I would focus on interrogating the technology they use. State-of-the-art machine learning relies on infrastructure that is quite unique. You should be able to tell from the infrastructure that they are using those approaches. Otherwise, they would not invest the considerable resources required to build it. It might be worth talking to their IT guy and ask questions that seem innocuous. Tell me about your server setup. Are you using the public cloud? Are you using Amazon Web Services? Are you using Spark (a framework used for efficient and scalable data processing of big data)? If you figure out their underlying technology stack, you can probably infer what sorts of techniques they are using. Nobody would set up a Spark cluster for running linear regressions.

What type of follow-up questions would you ask?

Why do you think that your approach is better? Have you benchmarked it to a simpler approach? How much alpha is created when you move from a single layer to your more complex model?

I have spent a lot of time finding disruptive, innovative companies. One of the things I have found is that you can't just listen to what the CEO is saying. The CEO is incentivized to overstate how innovative the company is to boost the stock price. It is more helpful to analyze the underlying investments that a firm that truly cares about innovation would be making. I have found several important factors that are important tells. Patents are obvious, but human capital is also very important and interesting. You can figure out who firms are hiring and what jobs are posted. Are they looking for someone with PyTorch experience? Are they hiring AI engineers? There could be two companies that both say that they are investing in Artificial Intelligence, but one is just saying it and not really doing anything whereas the other is spending top dollars on engineers from top tech firms.

I bet you could do something similar with hedge funds. The problem is that it might not really work on the smaller funds, but it could be something to try.

Typically, investors stay away from black boxes. They want to understand what they are investing in and what factors are important. One of the potential issues with machine learning approaches is lack of interpretability. What can hedge fund managers do to address that concern?

That's a very important issue. The standard approach for dealing with this issue is to evaluate feature importance. For example, you can permute a model by dropping out a feature to see how the performance degrades. You end up with a bar chart that lines up the features from the most important to the least important. For example, you might find price-to-book is very important for your model whereas 12-month past return is not.

There is another technique called LIME, which stands for local interpretable model-agnostic explanations. If you are using deep learning models or more complex models in general, you can use so-called surrogates, interpretable models such as linear regressions, to fit your more complex models with the goal of understanding better how the complex model works. Then you can evaluate the coefficients of the surrogate model and see what's going on.

What are some of the promising machine learning approaches that you are currently working on?

There is only so much value in using the data that has already been heavily analyzed. I don't want to discourage other people but think about how long Center for Research in Security Prices (CRSP) and COMPUSTAT (a popular database of financial, statistical, and market information on global companies) have been around. Do I really think that I can find something that nobody has been able to find before because I have a better statistical model? I don't see as much promise in this approach.

The situation is very different when you work with unstructured data. First, the technology is relatively recent. For example, BERT was developed in 2018. There hasn't been as much time to explore this space. Second, unstructured data is so much higher dimensional. Even if you and I train a deep learning model on the same text dataset, we can come up with two different answers. There is room for both of us to make money. Finally, unstructured data is growing exponentially. Currently, it is about 80 percent of total data. It is growing at a much faster rate than structured data. You have all this new information coming into play, it's brand new, and it hasn't been tortured yet by economists. I believe it's a fertile ground to be working on.

My advice for somebody in the hedge fund world who is looking for new research projects is to stop trying to squeeze out more juice from the structured datasets and instead get their hands on alternative unstructured datasets, get adept at the techniques, and start playing around. Try to look at things that people haven't looked at before.

I try to approach research projects with intuition. I come up with a hypothesis first. For example, I wrote a paper on brand. My hypothesis was that strong brands should outperform. How do you define a strong brand? I found the paper by Jennifer Aaker from Stanford who came up with the brand personality framework with five factors. The factors are sincerity, excitement, competence, sophistication, and ruggedness. I decided to use that framework and test the hypothesis that they mattered. Given the rubric that the five factors were important, I built a model and estimated loadings of each company on each of those factors.

Even with the complex techniques, the data is so high dimensional. You can't tell the model to go and find something interesting. It's not going to work. Instead, you need to utilize your own domain expertise to form hypotheses, such as that companies with good cultures outperform. Then you can come up with some factors that might explain the culture.

Once you do that, you have narrowed your search so much that it becomes a tractable problem for machine to solve.

Do you have any advice for someone who wants to become an expert in machine learning?

If you want to be effective in machine learning, you have to be full stack. Don't be the guy who shows up at the last second and takes all the glory because he is good at scikit-learn. Be the guy who takes the raw data in, is involved in actual scrubbing and normalization of the data, is intricately aware of how the database is built, understands the ETL (Extract-Transform-Load) process, and understands how the data is stored. If you want to be good at what you do, you have to be competent at every single step. Otherwise, you end up with a lack of context. I don't care how good you are at any part of the stack. If you don't understand the rest, you will run into problems at some point.

Do you think diversity leads to better results? Do you have any stories to share?

There are lots of different types of diversity. Let's talk about thought diversity, which I believe is crucial. It's similar to building a portfolio. If you try to build a well-diversified portfolio from 10 assets that are 100 percent correlated, it makes no sense. It's the same as having one asset. You want to find assets that have low correlation to your portfolio. Similarly, if you have a company with people who all think the same way, you should be looking to hire people who will break the group think, bring fresh ideas, and challenge the status quo. Such people are hard to find because most of the time people get promoted when they fit in and end up departing if they don't. If I were an HR person in a large firm, I would probably use unstructured data in resumes and build a model to select candidates.

As you know, this book is about quantitative hedge fund investing. What advice would you give to startup hedge fund managers and investors?

My advice to aspiring hedge fund managers is to be long-term oriented. Your long-term performance is all that matters. Remember that Warren Buffett earned 95 percent of his wealth after age 65. Success is about staying in the game over a long period of time. I've seen many funds that have big, splashy launches but in exchange have a very short runway because of the high-cost structure. This makes you fragile—even the best investors have down years. In order to maximize your chance of long-term success, you need financial resources, patient investors, and a deep commitment to weather the inevitable valleys.

My advice to investors is to be willing to take some career risk. There are a lot of studies that show that smaller, emerging managers outperform. Unfortunately, the industry is very institutionalized with a lot of career risk. However, assets under management are the enemy of returns. Some of the most innovative strategies I have seen only work with a smaller amount of assets under management. One example is crypto hedge funds. It is a brand-new space. It is not institutional. People are still trying to figure it out. The returns of crypto funds have been phenomenal. An investor who is willing to take some career risk can really benefit from considering unique firms and people with non-traditional backgrounds.

7.3 EMERGING MARKETS AND SUSTAINABILITY INVESTING WITH ASHA MEHTA

Asha Mehta, CFA, is Managing Partner & CIO at Global Delta Capital. Her thematic focus includes Emerging & Frontier Markets and Sustainability Investing. In prior roles, Asha was Lead Portfolio Manager and Director of Responsible Investing at Acadian Asset Management as well as an investment banker at Goldman Sachs. Early in her career, she conducted microfinance lending in India.

Asha was named one of the Top 10 Women in Asset Management by Money Management Executive and was profiled as a "Brilliant Quant" by Forbes magazine. She is a frequent speaker at industry conferences and her work has been featured in Pensions & Investments, the FT, CNN, WSJ, and other publications.

Asha is a Board Member of CFA Society Boston and an Advisor to the High Meadows Institute. In addition, she is an active advocate of financial literacy and financial empowerment. She is a supporter of several related organizations, including Compass Working Capital and 100 Women in Finance.

Asha holds an MBA with Honors from The Wharton School (University of Pennsylvania) and a BS, Biological Sciences and AB, Anthropology, from Stanford University. Asha has traveled to over 80 countries and lived in six.

Why did you decide to go into finance?

My parents grew up as refugees. My dad was born in the part of India that is now Pakistan. Once the British left, India separated, and that created war in that part of the world. My dad's family moved to India where they lived as refugees for almost twenty years. My mom's family was Jewish. They had to leave Europe during World War II. Both of my parents experienced extreme poverty and war. They strongly felt that medicine and education were the path to stability. Both of my parents became doctors.

I was raised in an academic environment, went to Stanford, and had a lot of pressure from my family to become a doctor. My plan was to go into medicine, and then I had an epiphany that led me into finance. I was working on my undergrad at Stanford majoring in biology and anthropology. I went to India one summer on a rural health project. I was inspired by my parents' stories and wanted to do something to help alleviate poverty. I was planning to be a doctor. I went to India to distribute vaccines in rural areas.

When I got there, my funding for the vaccine distribution project fell through. I was there for a summer staying with another family without a lot to do. The head of the household was a doctor. He invited me to join him to learn about healthcare in India. However, as I was trying to figure out how to be of service to the poor people in the rural areas, what hit me is the realization that the people there don't need a better healthcare system, they need more wealth, they need access to capital. They need funding to develop pathways out of poverty.

I didn't know what to do because I grew up in an academic household, but I knew that I wanted to help. It was before the internet. I got a phone book and started looking at the banks. I found the Bankers Institute of Rural Development. I thought it sounded interesting. I didn't have a phone. I took a rickshaw, I went to that place, and met the people running it. I told them I was in India for the summer and asked whether I could do an internship with them. It turned out to be a microfinance institution. It was in the late 1990s, and the microfinance was just taking off. It was before Muhhammad Yunus won the Nobel Peace Prize in 2006 for pioneering the concept of microfinance. I loved it. It was the most basic form of funding. I was going into villages and giving them one or two dollar loans so that women could buy a sewing machine, develop their businesses, and become entrepreneurs, and feed their families. That's when I decided to go into finance.

I went back to Stanford, finished my senior year with the focus in finance and infrastructure development. After Stanford I went to Goldman Sachs, joined their investment banking group that specialized in energy and power, right into the heart of infrastructure financing.

Asha, you have been very successful. You were profiled as a "Brilliant Quant" by Forbes magazine, you were named one of the Top 10 Women in Asset Management by Money Management Executive. However, I suspect that your career path has not always been easy. What are some of the obstacles that you had to overcome along the way?

I have been working in the industry for about 20 years. It is a dynamic industry, and change has brought both opportunities and risks. There are four obstacles that come to mind.

The first one was breaking into this industry itself. It is challenging to break into the buy-side, as it is competitive. However, the industry is fragmented, and a candidate can use this to an advantage. It is important to understand to find a niche where you can have an edge, add value, and be attractive from a recruiting standpoint.

The second obstacle was related to the fact that most roles that I have found interesting were in the quantitative space. Today, becoming a quant is a no-brainer. We live in the era of big data and machine learning; technologies have become accessible and commoditized. Most people use technology today, but 15 years ago it was novel. But I was recruiting in the early 2000s, just out of my Wharton Business School MBA program. There was nobody else in my class who was going into quantitative finance. Anybody who wanted to be on the buy side was going into fundamental roles. When I said I was going to a quant shop, my classmates looked at me as if I was crazy. They would ask me whether I wanted to be a programmer and questioned whether I would be compensated well enough. The role was too compelling to turn down, so I became a quant anyways. A lesson learned is that change happens, technologies evolve, and it is important to adjust to remain fresh and relevant. It is important to be willing to go outside of the consensus view and follow evolving thematics.

The third obstacle has been changing thematics. One thing that is really difficult on the buy side that requires stock picking and asset class selection is knowing when to stay the course and when to recognize that something has changed. There are changing thematics. If you are willing to take the risk and to be a first mover, that creates a lot of opportunities. I had two main challenges. I have already talked about my deep interest in international development. One of the first funds I managed was a frontier markets investment strategy. I thought it was fascinating to apply big data concepts to the most remote pockets of the world using computers without having to be on the ground. I enjoyed the work, but it was a small asset class that meant little to the firm's business. I had to wrestle with the question of whether I wanted to take a role that was inconsequential within the business. I made the right decision, because it drove the firm's brand and developed my career. A similar example played out over the last decade: In 2009, I brought my prior firm to be the first quantitative manager to sign the Principles for Responsible Investment (PRI). At that point, sustainability was a fringe topic, perceived to be concessionary to returns and risky for a business. In the time since, sustainability has gone mainstream and risen as one of the few growth themes within the industry.

The last obstacle I highlight relates to the topic of diversity and inclusion. It is challenging to rise in a buy-side organization. It appears as if it should be easy because equities are marked to market daily, and returns are transparent. Therefore, professional growth should be meritocratic. However, there are only a few roles at the top of an organization. For any person within our industry, it is important to figure out where the person wants to go, how one wants to grow, and what success looks like. The statistics are stark. In medicine, over 50 percent of graduating doctors are women; in law, about 50 percent of graduating lawyers are women. In portfolio management, women represent only 10 percent. We might be seeing a small change in the last couple of years because of the focus on diversity and inclusion, but the 10 percent level was flat for the previous 20 years. Therefore, it is important to understand how to grow within an organization and the inherent biases, to find the right path and create a supportive environment. These are challenges, but with every challenge, there is a tremendous opportunity.

You are an active advocate of financial literacy and financial empowerment. What drives that passion?

Financial literacy is a pathway to dignity and equality, a higher living standard for the poor. Ultimately, it comes down to access to capital. Access to capital is a tremendous moderating force in my worldview. Financial empowerment is relevant both domestically and globally. In my emerging market investment strategy, a core thesis beyond driving alpha is about providing access to capital. I believe that companies building productive businesses and pathways for a more prosperous environment should be rewarded with access to capital.

Outside of work, I spend my time on boards of non-profit organizations. One such organization is Compass Working Capital. It serves people at the lowest income level within the U.S. Specifically, it provides financial literacy and helps build wealth through savings accounts. The thesis is that one of the fundamental pathways out of poverty is the asset base, how much money you have in your savings account, rather than income, how much you get paid. A strong asset base allows people to take more risks and to feel confident about their day-to-day spending. I also support 100 Women in Finance. Many of the issues that women at the low level of income experience are similar. Their challenges are to gain and retain access to capital.

You lived in six countries. I am sure that you know from your personal experience about the positive and negative sides of diversity. Do you think diversity leads to better performance? What can organizations do to maximize the value of diversity and inclusion?

Diversity matters! Intelligence of a person can be measured by the person's ability to hold two opposing views in their head simultaneously. It is a talent to see things from different perspectives. On the buy-side we try to drive alpha, which is about identifying opportunities that are out of consensus. By definition, our industry is rewarding investors who are taking an alternative view. I believe there is no other way to develop an alternative view than to hear many different perspectives.

Ultimately, it is about cognitive diversity, it is about approaching a challenge, an asset class, or a situation with a diverse set of views. We often talk about gender and ethnic diversity. I think it is important. When we talk about cognitive diversity, it comes with a risk of arguing that a team of white men is cognitively diverse. While that could be the case, it is more likely to have cognitive diversity when the team is diverse across gender and socioeconomic, educational, and cultural backgrounds. I think that the industry should consider diversity that is more granular than cognitive diversity despite the fact that ultimately cognitive diversity is the objective.

What can organizations do to take advantage of diverse teams? I think it's important to build diverse teams and think broadly about the recruitment process. It is important to recruit from alternative networks, hire people who are differentiated within your organization, and then leverage their networks. However, it's not only about the recruitment and the pipeline. It is also important to ensure that all voices are elevated, respected, and heard. It is also essential that people of all backgrounds have the same opportunities for promotion and the same retention standards are applied to all employees. That can be hard to do because it is easier to promote someone who thinks like you. It is important to be aware of your own biases.

As organizations grow and achieve a certain status, their maturity comes with legacy in infrastructure, technology, philosophy, and also ways of thinking. That creates opportunities for entrepreneurs. Our industry is so fragmented. When individuals don't have an opportunity to grow within an organization, they have an opportunity to build another organization. The cost of entrepreneurship today is lower than it has ever been. The cost of technology today is lower than it has ever been. It is relatively easy to build an organization from scratch, which makes it feasible for diverse teams to test your hypothesis that diverse teams outperform.

You specialize in emerging and frontier markets and sustainability investing. I know that you recently wrote a book about it. What can readers learn from your book about this style of investing? Why should hedge fund investors be interested in this investment segment? What's the best way to access it?

Emerging market investing and sustainable investing are two distinct topics. The book discusses how the emerging markets have fundamentally shifted in the last two decades. With a few exceptions, they have gone through significant advancements in terms of economic globalization. We have seen liberalism and economic liberalization bring substantial capital flows into emerging markets. As I have mentioned before, access to capital either at individual level or macro level fundamentally changes the opportunity. Access to capital, technology, and stronger governance and development programs has turned emerging markets not only into a compelling investment opportunity for investors but also a fundamentally different backdrop in the global landscape.

Twenty years ago, China was an impoverished country. Today the young people in China have never known a life of poverty. Today China is one of the leaders in technology development. India is another notable leader in technology development and education, that is essential for promoting future generates to continue advancing. Two decades ago, emerging markets were about oil. Today emerging markets lead technological development and have an incredibly rapidly growing middle class. The domestic demand theme has fundamentally shifted.

The sustainability aspects only advance this theme. As one invests with a sustainability lens, that creates greater consumer and labor classes and produces better performance. Sustainability as an asset class is in its early stages. People often think of sustainability from an old fashion exclusion-based SRI (socially responsible investing) perspective, which comes across as concessionary in returns. The thesis of my book and of my business is that it is not the case. Sustainability enables investors to get a broader view of how capital is utilized and distributed and how management teams are behaving, which gives better insights into investment opportunities.

Since emerging markets and sustainability investing are two distinct themes, combining them is additive from the performance perspective. The beta long-only opportunity in emerging and frontier markets is exciting. They don't have the same boom and bust because of better governance at the central bank level and lower sociopolitical risk. The growth in these markets exists at the macro level with emerging markets growing faster than developed markets, but it is also showing at the corporate level. That's important because macro opportunities don't always translate into market opportunities.

Over the last decade U.S. tech companies have outperformed all other stocks around the world. However, some emerging markets have kept up with the tech companies. One example is Saudi Arabia which has gone through a dramatic process of liberalization over the last decade. Although today the developed markets are tightening from the fiscal and monetary perspective, many parts of the emerging markets are seeing easing from the central banks. Therefore, with the valuations being depressed in the developed markets, the growth-relating themes in emerging markets are very compelling.

From the alpha perspective it is even more exciting. The emerging markets are still off the radar for many investors, both institutional and retail. Therefore, the securities are still inefficiently traded, which is fascinating. There are fragmented investors with fragmented objectives. It is a large enough asset class for which we can build robust models, but it is small enough for inefficiencies to exist. The opportunity for alpha generation within this market is very rich. The sector-related themes in equities are quite interesting. We don't have to be long all Saudi stocks, but maybe just Saudi consumer and telecom stocks. We can also talk about China bank and China tech stocks that are also quite unique.

I understand that emerging market investing and sustainable investing can be attractive. However, investors often look for a risk-managed approach and they might worry about political risk. How do you suggest applying quantitative approaches to address those concerns?

I believe that quantitative approaches are ideal for investing in emerging markets from the return and risk management perspective. We can generate alpha using quantitative approaches because the asset class is traded so inefficiently. As a quantitative manager, I can cover every single stock in more than 50 emerging and frontier market countries, a universe of around 20,000 securities. By contrast, a fundamental investor has to spend time with boots on the ground and can only cover 20 to 40 stocks at a time. Therefore, a fundamental investor can have a portfolio breadth, which is only a small fraction of the breadth available to a quantitative investor. That's a tremendous loss of opportunity, which is particularly significant because the inefficiency is so great. I see the emerging markets an order of magnitude less efficient than the developed markets. However, if you consider the small cap space in emerging markets, where a lot of the consumer-related themes play out, that part of the market is even another order of magnitude less efficient. It is hard to cover the same set of securities for fundamental investors. Thus, part of the strength of a quantitative process in emerging markets is the breadth itself.

The hurdles of investing in this asset class are huge. The transaction costs are very high. The liquidity and foreign ownership can be significant obstacles. However, as a quantitative investor with access to such a broad universe, I can easily substitute an attractive stock with a high hurdle with another attractive stock that doesn't have that hurdle. That's a huge advantage for quantitative investors.

Emerging markets are volatile. Quantitative investors can bring the discipline. For example, Pakistan sold off dramatically years ago on severely negative headlines. Quantitative investors have the discipline and the rules to separate headline noise from the changes in key fundamentals. Similarly, in the moments of euphoria, there is a sell discipline built in.

The return side or the opportunity can be considered in a multi-faceted way. It is important to differentiate between a multi-factor perspective and a multi-faceted perspective. The multi-factor approach is valid. Companies can be evaluated based on their valuations, growth prospects, and technical support for stock price. However, it is valuable to think even more broadly than just factors and consider a holistic set of inputs such as the macro environment, the transaction costs, and the impact, which is important for sustainability. This holistic set of inputs can be incorporated into a robust quantitative process.

7.4 KEY TAKEAWAYS

Following are the key takeaways from this chapter:

- Trend following is about applying mathematical techniques to measure where markets are going and what the markets are doing. Trend following has stood the test of time, and it has provided crisis alpha.
- Machine learning offers a powerful toolbox of methods but "ultimately, you have to pick the right tool for the job." Unstructured data offers many opportunities for creative applications of machine learning to trading.
- Access to capital, technology, and stronger governance and development programs have turned emerging markets not only into a compelling investment opportunity for investors but also a fundamentally different backdrop in the global landscape.

CHAPTER 8

Expert Hedge Fund Investors

"An investment in knowledge pays the best interest."—Benjamin Franklin.[a]

"The only source of knowledge is experience."—Albert Einstein.[b]

We continue our conversation with extraordinary individuals. This chapter includes personal stories of expert hedge fund investors. They talk about their career paths, openly discuss their exacting search for investment answers, and share invaluable insights for young people and seasoned investment professionals. It covers the challenging topics of quantitative and qualitative manager selection and quantitative portfolio construction.

8.1 QUANTITATIVE AND QUALITATIVE MANAGER SELECTION WITH ADAM DUNCAN

Adam is the Head of the Investment Science Unit and is a Partner and Managing Director in Cambridge Associates' Boston office. He is the former head of the Portfolio Modeling and Analytics Group where he oversaw the development of the factor models and portfolio

[a]Benjamin Franklin was one of the Founding Fathers of the United States, a drafter of the Declaration of Independence, and a polymath.

[b]Albert Einstein is regarded as one of the greatest physicists of all time. He was a Nobel Laureate in physics, 1921.

DOI: 10.1201/9781003175209-8

decomposition tools in use at Cambridge Associates. He has also held senior roles in manager research where he focused on investment opportunities in diversifying asset classes and strategies, including CTA/managed futures, alternative beta and risk premia, active currency, and volatility trading strategies as well as discretionary and systematic macro programs. Adam's research interests include skill and anomaly detection, predictive modeling related to manager selection, and process automation. Adam is a frequent presenter at industry conferences and guest lectures at a number of universities.

Prior to joining Cambridge Associates, Adam served as Director of the Global Currencies Client Risk Advisory Group at Credit Suisse where he helped clients achieve best execution and optimize their currency hedging programs. He conducted research on topics such as tail risk hedging, strategy selection under volatility regimes, and event risk positioning. Previously, he spent twelve years at JPMorgan Chase, where he held several roles focused on the trading, structuring, and sales of fixed income and foreign exchange derivatives.

Adam, as you know, I look up to you as someone very thoughtful about applying a rigorous scientific approach to answer challenging investment problems and to make solid investment decisions. You have an interesting background and career path. Could you please share the story of how you decided to go into finance?

Early on, I wanted to be a computer scientist. I was fascinated with computers as a kid. I loved working with computers. When I started doing my undergraduate in computer science at Carnegie Mellon University, I quickly found myself in a very advanced form of computer science. I was struggling. I had hit my limit cognitively. As a young person, I couldn't deal with an extremely high level of rigor. So, I switched my major from computer science to information and decision systems, a hybrid of computer science and game theory. Sometimes young people make choices like that, and I think they don't get enough guidance to hang in there and keep on track even though it's difficult. I wish I had a mentor who would've helped me think through those decisions.

Somehow I got the bug for finance. It seemed like something I could do. I also realized that I needed to be around money to make money, which was important to me then. So, I decided to leave computer science and started pursuing finance.

It was a challenging time in 1992. I joined a regional bank. I worked my way up to a trading desk, where I met my future wife. She was a brilliant repo trader. She trained me. We traded the short-term interest rate book together. I started as a Fed Funds trader. After some time, I worked my way up to trading interest rate derivatives and helped manage the bank's interest rate swap book. We were running a liability hedging book of derivatives and it was the heyday of derivatives. I loved it. After a couple of years, my wife and I decided to go to business school together at Carnegie Mellon University.

When I got there, I jumped right into the deep pool of mathematics and computer science. The finance classes were very quant heavy. I took every finance course I could: fixed income, option pricing, etc. It feels like I took forty-seven finance classes and maybe one marketing class that I had to take to graduate. I loved studying finance, optimization, and econometrics. I was fortunate to get exposed to brilliant professors. I still keep in touch with many of them and look up to them still.

After graduate school, I got an internship and ultimately a job at JPMorgan on the fixed income sales and trading team. First, I worked on fixed income structuring using swaps and other complicated derivatives. Then I transitioned to the credit hybrids area and worked on innovative products such as options on *credit default swaps* (CDS) and constant maturity CDS. A year later, I moved to the foreign exchange (FX) trading desk, covering the private bank for structured notes. We were taking complicated derivatives in foreign exchange, putting them into notes, and selling them to high net worth individuals. I loved working closely with the fixed income research team. We were finding many exciting opportunities in relative value—something that I still find very interesting. Relative value in fixed income and FX is a fruitful area of research, particularly for those with a computer science background. Programmatically searching spaces, surfaces, and curves with specific ideas in mind is a fantastic way to marry computer science and finance. It also forces you to be very explicit about what you are doing. Expressing yourself in code is very liberating in this way.

My Wall Street experience was an incredible time of my life. It was extremely challenging but also very exciting. I worked on many interesting projects. I worked extremely hard, but I was also very stressed because of the nature of the industry and working in New York. I have also seen some challenging market environments. For example, I was on the trading desk on Sept 11th and during the Global Financial Crisis of 2008. All told, I spent about 11–12 years at JPMorgan, and then I spent a little more time doing similar work at Credit Suisse. When my group at Credit Suisse was terminated, I decided that I wanted to do research. This was something that I had wanted to do for a long time and was borne out of my experience with the fixed income research team at JPMorgan where I had many great mentors and collaborators. Terry Belton, Gagan Singh, and Srini Ramaswamy had major influences on me.

It was not an easy transition to research since my background was in sales and trading. Fortunately, I got a call from Cambridge Associates. They were looking for a macro researcher in their hedge fund research area. Twenty-six interviews later, I got the job and transitioned into a completely different area of finance. I soon realized there was a whole world of investing that was completely detached from the trading desk and my Wall Street experiences. And I knew almost nothing about it.

Looking back, I recognize now that my experience on the trading desk helped me a lot later on in my career. It's sad that when you are on the trading desk, after a while, you develop a sense that there is nothing else out there that you can do. That's so not true. There is so much more to the world of finance that people on the trading desks are unaware of. For example, I knew nothing about endowments and foundations, pensions, or hedge funds. I knew hedge funds as counterparties to my trades, but I didn't know anything about manager selection. And I knew even less about the world of private investing.

You've accomplished a lot to become the head of investment science at Cambridge Associates, but your journey was not always easy. You already mentioned your early challenges with advanced computer science school work and the heavy stress level while working on the trading desk. What do you think is the biggest obstacle that you had to overcome along the way?

One of the biggest challenges for me has been being in the middle between sales and trading. The world tends to reward specialization. People who specialize and are very good at an activity, get rewards and earn rents for that. I was one of those people in the middle, which posed a set of unique challenges along the way.

Generally speaking, when you are on Wall Street, you are either in charge of client relationships or you are in charge of the risks in the trading book. If you are like me, who was a structurer or a designer of solutions, you are in the middle trying to come up with clever things to put in front of clients and do it in a way that is palatable for the traders and risk takers. That's a very tenuous position to be in. When times get tough, the relationship people can lean on the strength of the client relationships they have developed, and the risk people can lean on their skills managing the complicated risks in these trading books. The folks in the middle, like solutions experts, are often left with little to lean on. I often wasn't the primary relationship person or the primary risk taker.

Another challenge was that I didn't have a Ph.D., I only had a Masters degree. Not having a Ph.D. shut off some of the research pathways I wanted to pursue. It's hard to get on the research staff because it doesn't carry enough cachet for the firm to have a non-Ph.D. putting out important content. Once again, I found myself in the middle-ground between undergraduates and Ph.Ds. When you find yourself in a middle ground, it can be challenging to carve out a place where you can be comfortable and be successful.

That was probably the most challenging thing for me—being in the middle ground, not fully embedded with the clients, not being fully embedded with the risk takers, and not having a Ph.D.

Adam, you are one of the most sophisticated investment professionals I know. Your title of the head of investment science perfectly matches you. It sounds crazy to me that you couldn't pursue certain research areas because you didn't have a Ph.D. You constantly read academic papers, evaluate proposed approaches, come up with your own techniques. How would you compare academia and the industry? What is the gap between them? What can practitioners learn from academia? What can academics learn from the industry?

That is a great question. I used to be very skeptical of academia. If you consider how many Ph.Ds. are minted each year, and read the papers that come out, you don't see massive contributions to the canon of understanding in finance. Of course, most of the time these contributions are very marginal.

One of my former professors and someone I respect immensely told me once: "Don't be too hard on the academics. We would have a lot fewer good ideas, if it weren't for them." I think he is exactly right. It's the ideas, not the lack of realism in the assumptions, or that it doesn't hold up to empirical evidence or that maybe the researcher fell victim to *data mining* or whatever.

It's the ideas that are important. It's the frameworks and mental models that are super useful for organizing your thoughts about complicated concepts. The world is an incredibly complicated and noisy place. These abstractions are necessary and extremely helpful. When people write down the math and suggest a framework for explaining something that is very complicated and noisy, it's extremely useful.

For example, in my personal experience with asset pricing models in studying the cross-section of hedge fund managers, I find that the academic models, such as the Fama-French three factor model, are directionally correct. I find that the models are not completely descriptive, but they are mostly descriptive, and the empirical evidence lines up reasonably well with the models. In this way, academic models provide a very useful way to view the world. Of course, we can argue about the validity of individual factors, in-sample vs. out-of-sample effectiveness and all of those things, but overall we get a lot of very practical and useful ideas from these models and academia generally. I like paying attention to academia because they are responding to questions from the industry and are doing so in a really competitive space. This mostly results in good outcomes for practitioners.

For their part, industry practitioners provide a great forcing function on academia because academia tends to take up the topics that matter most to practitioners. Practitioners explain the challenges they are facing and in some ways define the problem. Academics listen and bring their best thinking to these hard problems. It's quite a useful and productive symbiotic relationship.

It's very rare that academic approaches fit neatly into the practitioner's specific problem. For example, machine learning is a popular topic. If you go and study machine learning, there is a pretty good chance that you are going to learn about machine learning in the context of image identification. This is true for a number of reasons, but that was the industry's most proximate use case. There are a lot of images available, lots of interesting problems in computer vision, and the available models tended to work well on that data. When you go back to a specific finance task, you have a model that can identify airplanes and automobiles with incredible out-of-sample accuracy, but it has little to say directly about how to solve your problem. The ability to go from what you've learned about image classification to building a model that can help select hedge funds is an important skill that practitioners need. The real skill is looking at what is done in medicine, image classification and many other domains, and then figuring out which nuggets you can take away to help solve your own problems.

Manager selection is one of the most challenging problems in asset management. I know that you've personally evaluated most manager selection techniques. In my opinion, you are one of the top experts in the world in this area. Could you please tell us which quantitative hedge fund selection approaches are worth considering?

I love this question. That's one of the reasons my group exists. The question your boss is going to ask you is some like: "Out of these 100 funds, which funds will outperform their benchmarks out-of-sample over the next five years?" The problem with that question is that it's a Nobel prize worthy question. Meaning, if we figure that out, it's likely we will win the Nobel prize in economics. But like many problems in finance, it is an incredibly noisy problem. We recognize this class of incredibly difficult problems (we call them "Nobel Prize Problems") and try to transform those questions into something that we can solve. This is the real art of applying quantitative techniques and machine learning to manager selection.

In manager selection, typically people jump directly into the analysis of track records and performance metrics. They say: "I have a huge database, lots of track records, let me see if I can extract something and build a selection model." That's an extremely difficult problem. I don't want to say that you cannot get any signal out of it, but the amount of signal you get is going to be very low. It doesn't really matter what financial metrics you use because I think I've tried them all. It is my view that the reason you cannot find signal in these track records is because the answer is not embedded in the track records to begin with. My view is that these track records are largely informationless. In other words, they don't have much information about the underlying investment and its prospects for future success. Thus, mining and reshaping track records with different sorting methods or applying different mutations is really just moving the randomness around. You really aren't getting any closer to something that's usable for predicting expected returns.

Where does that leave you? What can you do? I think the answer is going back to the process that you are trying to model in the first place. Keep in mind, Cambridge Associates has been doing manager selection for decades with good success. And that process is largely not about analyzing track records.

A good first step is extensively interviewing your research staff and writing down what they say. You will learn a great many things. As you do that over more and more researchers, you will start seeing commonalities in the things that they look at and the approaches they take.

A next good step is to use that information to map the process to data. As with anything, some things are very difficult to map to data and other things are quite straightforward. When you follow this process, you start to build a model that behaves like your researchers behave. Presuming that your researchers do a good job of picking managers and you have a reasonable data story to support it, you have a chance of building a model that replicates researcher behavior. We have had good success with this process.

If you already have a great team of researchers, why even bother trying to build a model?

When I used to do manager research, I often looked at my list and the managers I recommended, and asked the question: "How did I find these managers? How did fund A or fund B end up on my list?" Was it that I had a list of characteristics I cared about, codified that list of characteristics, filtered some large database systematically to then get a subset of funds, interviewed each manager from the subset, and A rose to the top? Or did I just mine my social networks and other interactions to develop relationships and subsequently conviction in the managers? It was certainly not the former.

To me, this was a disturbing way to approach the world if you are a quant. A quant doesn't want to randomly traverse the addressable universe through conferences and the like. We want to do the former process I described. One of the things I always worried about when I picked a fund was all the other funds that I didn't look at. How do I know that I am seeing everything? For example, I know that databases have hundreds of Commodity Trading Advisors. I didn't look at all of them before I made my choice. What if I was reaching some "local maximum" so-to-speak?

Because of the labor intensive manual nature of the process of manager selection, you will always be worried that you are not seeing the whole universe of available funds. By codifying the process and knowing that it behaves a lot like you behave, we can apply this model to a large database of funds and give researchers some comfort that they are seeing everything before making selections.

Adam, it sounds as if you are saying the following at a high level. Track records are largely informationless. There are some researchers who are good at identifying relevant factors and selecting superior managers. However, their work is not scalable because it is so labor intensive. A model that replicates that process can identify a broader pool of good managers. Is that what you are trying to accomplish?

That's correct. I can create a target-rich environment for researchers to survey so that instead of a database with 10,000 funds they can get a list of 75 funds that fit with their preferences and research process. That pool of 75 funds should be very hard for the researcher to reject easily. Those funds survived a many-featured winnowing process that matches the researchers own style and preference. This is a significant evolution in how we use data to help our researchers identify high quality ideas that are consistent with their process.

Adam, thank you for sharing your insights on manager selection. I want to switch gears and talk about diversity and inclusion. Do you think there are some benefits of diversity in hedge fund investing? What is your personal view? Have you found any empirical evidence that shows the benefits of diversity?

My team is very passionate about this topic. I have a very diverse team, and we believe strongly that diverse teams lead to superior outcomes. Cambridge Associates just hired a new head of diversity to guide us along this path to not only improve our approach to diversity, equity, and inclusion but also to use our position in the industry to increase diversity in the industry.

Currently, people are largely mining their own social networks to find diverse managers and make sure that these funds are getting in front of research teams. That's good when you don't have anything else to lean on, but our question is "What is the data story here? Is there anything we can be doing at the strategic level with data to help more diverse managers to get exposed?" Also, "Is there a systematic way to create 'diverse/not-diverse' labels for every manager in a database?"

We think there is a way to systematically identify and label diverse/non-diverse managers. It is a complicated process involving multiple points of data ingest that must each be validated and then harmonized into a single label. Hardening this process into something that is robust and lasting is something we are working hard to achieve.

There are two potential benefits of diversity. The first one is a societal benefit. We know it's the right thing to do to embrace diversity. But the second potential benefit is improving investment performance. Are there any performance benefits of investing in diverse teams or managers?

This is a really tough question to answer at the moment. At a very first principles level, we are still working through the process of assigning diversity labels to the dataset. Without a robust labeling process where we have faith that diverse managers are identified correctly, it's pretty much impossible to step up to the performance impact question. As you know from the factor space, definitions are often unclear and methods can vary for what seem like pretty straightforward data elements. The ESG space is really struggling with this specific problem right now. The lack of unifying standards from which baselines can be established and measured is complicating things in the ESG space right now. I think we'll get through that, but it will take some time.

Separating the effect of diversity, after controlling for all the other confounders, is an impossible task right now. There is almost no way you are going to convince me today that you have controlled for everything correctly to draw a causal link between diversity and performance. It might even be difficult for you to convince me that you even have the dataset labeled consistently and correctly.

As you know, this book is about quantitative hedge fund investing. What advice would you give to hedge fund managers and investors that can help them improve their performance?

Here is an observation from manager selection modeling. Suppose I interview you, write down your investment process, capture a lot of the dimensions of your investment process, identify many features that you care about, apply those features to a large database, return all the funds that match your criteria. Suppose further that the list we get back is very different from your current recommended list. What should I conclude about your process? It's either what you really care about is different from what you have told me or perhaps you make a lot of exceptions. When people say, "use systematic methods to make performance go up", what I hear is, "let's take this already good process and see if we can make it more repeatable." I think the real performance gains come less from some magical feature you uncover and more from just implementing your current process with more discipline. Doing diagnostics like this and figuring out things like, "why isn't my current list reflected by this model?" is a super valuable activity and where I would look for performance gains.

It's a great opportunity for learning and it's one of the steps on a migration toward a more systematic and scaleable approach to manager selection.

Another possible conclusion is that what you are not picking up in the data is really important. For example, we cannot pick up in the data that two portfolio managers are not getting along with each other and it's causing problems for the fund or that there is an internal power struggle that is very distracting for the investment team. This is where the power of the human researcher really shines. "Human-in-the-loop" models like this can be incredibly powerful. There are still lots of things that we struggle to put into a model that a researcher can know firsthand.

There are a few other interesting things. For example, if you tell me all the things you care about and we define the features of the funds, we apply those to a database, and I show you the results—a list of 75 funds. However, if I cover up the names and ask you to go through the list to select Yes/No/Maybe, it becomes very disorienting. Your ability to index on the names influences how you label those 75 funds as good or bad ideas. This is a form of halo bias. The fact that you know it's AQR or PIMCO (Pacific Investment Management Company) influences how you respond to the model output. Try it for yourself!

Most young people who try to apply machine learning in finance jump right into the data and the track records. I might recommend doing the opposite. Start with interviewing your researchers and try to solve the problem with simple manual heuristics first. Have your researchers manually label the outputs from the model as good or bad and put that feedback back into your model. Eventually, you will arrive at something that works well and was built from a deep understanding of the human process. Then you can begin to train a model that can learn some of these deeper, non-linear aspects of the process. We spend a lot more time interviewing our researchers than we do tuning model hyperparameters, that is for sure. And watch out for those Nobel prize problems. You can waste a lot of time trying to wring a tiny amount of signal out of these really noisy datasets.

8.2 QUANTITATIVE PORTFOLIO MANAGEMENT WITH CHRISTOPHE L'AHELEC

Christophe L'Ahelec, Senior Director of external manager program at the University Pension Plan (UPP) Ontario, is an investment professional with more than twenty years of global experience in France, Hong Kong, Switzerland, and Canada. He works with the UPP team to shape UPP's global investment strategy with a particular focus on its External Manager Program. He is building strong trusted relationships with internal and external stakeholders to generate the risk/return objectives required to strengthen the future of UPP defined benefit pension plan for generations of university employees to come.

Prior to that Christophe was a Senior Principal in the Capital Markets (CM) department, at Ontario Teachers' Pension Plan Board where he was accountable for the top-down management of the external managers portfolio and was part of the investment committee allocating to hedge fund, private credit and long-only equity managers. During his 11-year tenure at OTPP, he originated, conducted due-diligence, and on-boarded systematic alpha managers in the equity and futures space, along with leading the group's work in the alternative risk premia space.

Before coming to Canada, Christophe was a quantitative analyst and assistant portfolio manager at Mignon Geneve SA/Alpstar Asset Management, in Switzerland, where he took part in the creation and portfolio management of a European systematic market neutral fund. He spent the first two years of his career working for formed by a merger of Banque Nationale de Paris (BNP, "National Bank of Paris") and Paribas. in Hong Kong where he provided real-time quantitative support to the fixed-income trading desk.

L'Ahelec is a graduate engineer in Finance and Applied Mathematics from the Ecole Nationale Superieure d'Informatique et de Mathematiques Appliquees de Grenoble, France and holds the Chartered Financial Analyst designation. He has been invited to speak globally on advanced portfolio and risk management topics and is also known for his work as an academic writer. His papers have been published in the Journal of Asset Management and the Journal of Alternative assets.

Why did you decide to go into finance?

I am not one of those people who wanted to go into finance as a kid. I grew up in France. Even when I was in high school, I was unsure about what I wanted to do. Instead, I knew what I did not want to do. I did not want to be a doctor or a lawyer. I come from a family of engineers, but I did not want to work in the metallurgy, paper, or the automotive industry like my father, my grandfather, and my great grandfather. I was good at sciences. After high school, I took what is called La Voie Royale, or the Royal Path, a program that included two-three years of intensive undergraduate-level training in advanced mathematics and physics. At the end of the program, students take competitive exams and earn acceptance to engineering schools based on the relative ranking. Around that time, I started getting interested in financial markets and I chose to go to Ensimag (Ecole nationale superieure d'informatique et de mathematiques appliquees de Grenoble), a prestigious French Grande Ecole located in Grenoble. The school is most popular for its computer science program, but it also has an excellent applied mathematics program that includes financial engineering. I decided to go to Ensimag because of the financial engineering program.

I took many computer science classes. I did not find them particularly interesting then, but today I regret that I did not take them more seriously. Instead, I really enjoyed studying probabilities, statistics, stochastic calculus, and financial engineering for three years. During the summer after the first year, I did a one-month internship at a small trading room of a regional subsidiary of the Societe Generale in Strasbourg. It is there that I discovered the foreign exchange and fixed income trading. During the second summer, I spent one month in the fixed income trading room of Banque de France, the Central Bank of France, in Paris. We managed the Treasuries and other investments of the bank. During the third year I did a compulsory long-term internship. I spent six months in the fixed income financial engineering department of BNP Paribas in Paris.

So, I got into finance as a quant. I like looking at investments from a quantitative perspective. I also like that it is an ever-changing world. The world today is very different from twenty years ago when I started my career. You always have to innovate to stay afloat.

Christophe, you have a very impressive resume contributing to success of large and medium size pension plans. Could you please tell us about your journey? What were some of the obstacles that you had to overcome along the way?

I have had a weird journey looking back, but I believe everything happens for a reason. When I finished the internship at BNP Paribas, they offered me a contract to work in their fixed income trading group in Hong Kong. At the time, I didn't really plan to live abroad. I grew up in France, I travelled in Europe, but never lived overseas. However, I chose to accept the offer because I thought it would be nice to have experience working abroad, especially in Hong Kong, and it would be a great opportunity to improve my English, which was terrible then. My grade on the final English exam at Ensimag was the second worst out of everyone.

The first few months in Hong Kong were very challenging because I didn't understand half of the things that I was told. On the first day at BNP Paribas, I had to watch a two-hour video on anti-money laundering, and I didn't get most of it. When I had to open a bank account at HSBC, my friends told me what to ask for, which was fortunate as I didn't fully understand the documents I was signing.

Initially, it was supposed to be a fifteen-month term, but I ended up staying there for twenty-two months because I quickly realized that I didn't want to go back to France. Hong Kong is a multi-cultural international city. I loved living abroad, discovering new culture, working with people from other countries. I fell in love with Asia.

My role within the fixed income trading room was support and development of front office applications. After about a year, I wanted to move to a portfolio management role. It was around 2001–2002, not a great time to look for a job. I kept looking for trading opportunities, especially in Asia, but could not find anything. At the end of my contract term, I went back to Europe and after a few months I joined a hedge fund in Geneva, Switzerland. Although Geneva is outside of France, it is still very similar to France. The firm used to be a fund-of-funds, but in 2002 the founder decided to expand the business to hedge fund trading strategies, specifically credit and equities. He built one team on the credit side and three teams on the equity side. One team focused on equity long/short, fundamental discretionary trading. Another team applied macro approaches to trading. My equity team developed systematic market-neutral strategies.

My team was responsible for launching a fund in 2005. We had to build all the operational capabilities from scratch. Data management, middle office, back office, and risk management capabilities. I designed all risk reporting for all the equity funds. At the top, my team managed about 130 million Euros in this systematic strategy.

In 2008 I decided to leave the firm and move to Canada. My wife is from Japan. She and I didn't particularly like living in Switzerland. She didn't want to go to Asia, and I didn't want to stay in Europe.

We considered the U.S. and Canada but chose Canada because it was easier to immigrate there. And we had Canadian friends who sold us Canada very well! We applied for permanent residency and moved to Canada in 2009 in the middle of the Global Financial Crisis with a lot of uncertainty and without any job opportunities. I feel like I am looking for new job opportunities always at the worst times!

I started actively looking for a job right away. I had tried doing that from Geneva, but it was very hard to do that remotely. I got very lucky because I landed a role at the Ontario Teachers' Pension Plan (OTPP) within a few weeks. At the time I didn't know much about the pension plan industry. I didn't know anything about the OTPP.

It all happened by chance. I don't think I would've applied for the job if I saw it on the OTPP website. I was talking to a headhunter about risk management jobs. After about five minutes, a partner from the recruiting firm walked in, introduced himself, and told me: "Christophe, forget about this risk management role. I have a perfect job for you!" He showed me the job description for a position at the OTPP. I could see myself as a good fit, but I didn't meet all the requirements. For example, they required macroeconomic expertise, which I didn't have. That's why I don't think I would've applied. The headhunter told me: "Christophe, there is the job description, but I also know what they are looking for. I think you are the perfect fit." He was right. After four or five interviews, I got a job offer and spent almost eleven years there.

It was a great journey. I learned a lot. I worked with fantastic, very smart, and innovative people. However, there are cycles in everyone's careers. After eleven years I was ready for the next challenge. In March 2021, I joined the University Pension Plan Ontario, a new pension plan that officially launched on July 1st, 2021. The mandate of the UPP is to consolidate pension plans of universities across Ontario. We launched with three founding universities. A very different setting than the OTPP. The OTPP is a big established pension plan managing 250 billion Canadian dollars, with 1,300 employees across three different offices, an investment team of 350 people. When you are one of 350 people, the impact you can have on the success of the investment strategy and the investment areas that you can cover are very different compared to when you are on an investment team of ten, twenty, or even thirty people. The UPP currently has six people on the investment team. Although my main responsibility is heading the external management program, I am also involved in all aspects of investment business such as asset allocation, cash management, and rebalancing. We are setting up a pension plan from scratch while also managing the three portfolios that we have inherited from the three universities. It is a very exciting opportunity!

You have a very strong academic background. I was fortunate to write several academic papers with you. How would you compare academia and the industry? What is the gap between them? What can practitioners learn from academia? What can academics learn from the industry?

Sometimes there is a gap or disconnect between academia and the investment industry. Sometimes academics lack real investment experience and market knowledge. They make assumptions that don't reflect the real world and, therefore, come to conclusions that are invalid and cannot be justified from a practical point of view. When I worked at the hedge fund in Geneva, we were looking to hire someone with a strong quant background. We interviewed a guy with a strong scientific background with a Ph.D. from one of the best engineering schools in France. We asked him to tell us about his Ph.D. research. He spent twenty minutes telling us about his research and writing complex mathematical formulas on the white board. None of us on the team understood a bit of what he was saying, but the three of us ended up sharing the same conviction that his research was impractical, and it could not be applied to the real world.

On the other hand, academia can bring a different perspective to a problem that a practitioner can be facing and suggest a different approach that a practitioner can then explore. Sometimes academic approaches with some adjustments can be applied in the real world. Practitioners can learn a lot from academia.

I want to be careful about overstating the disconnect. For example, there are organizations such as the Edhec Risk Institute that can successfully marry the two worlds of the academia and the industry. They look at the investment world with academic lenses while keeping feet on the ground and making practical advances that can be used in the investment industry.

One of the big challenges that I see in the industry is lack of clean datasets that are sufficiently large to draw statistically significant conclusions. Academics often use large datasets that are too long to be reflective of where the world is today. For example, the market structure today is very different from the market structure ten, twenty, or thirty years ago. The market participants have changed. Because of that it is not always helpful to rely on outdated data. Since you may need a hundred years of data for statistical significance, academics may completely dismiss some of the work done by practitioners. However, I think this type of analysis can be an important source of information for making decisions going forward. Of course, it requires prior beliefs about the world that are based on practical experience.

There is a lot of noise in the financial data, but the world is also very dynamic. A practitioner has to decide what data is relevant and use it even though it might be a relatively small dataset. The practitioner can still draw useful conclusions when combined with the previous experience and judgement even though that doesn't make as much sense from a purely statistical perspective that ignores prior beliefs. It is more consistent with a Bayesian framework.

At Ontario Teachers' Pension Plan, which is one of the most sophisticated pension plans in the world, you were responsible for the top-down portfolio management of a $26 billion external managers program allocated across hedge funds, private credit, and long-only equity managers. What is your approach to portfolio construction?

My portfolio construction contribution was more on the hedge fund side of the portfolio. The OTPP is definitely a sophisticated and innovative firm, but our approach to portfolio construction was not very sophisticated from a quantitative perspective. We didn't rely on any optimizations because we didn't believe in them. I am not a fan of any optimization algorithm or a system. They can inform your portfolio construction process, but I would not blindly follow any optimization system.

We used a bottom-up approach to portfolio construction combing both qualitative and quantitative factors while also controlling for top-down diversification and balance across asset classes, strategies. For example, we tried to balance the convergent strategies such as relative value and M&A (mergers and acquisitions) and divergent strategies such as trend following or global macro.

The hedge fund portfolio mostly included uncorrelated alpha strategies. Of course, it is inevitable to have some beta in certain parts of the portfolio, for instance, on the credit side. However, for the most part we allocated across uncorrelated alpha strategies. We regularly compared our discretionary approach to several systematic portfolio construction approaches such as an equal risk contribution or an equally volatility weighted approach. We found that our approach was very similar to both risk parity approaches subject to some additional considerations such as manager capacity and non-investment risks.

I have also learned while at OTPP that it is very challenging to time hedge fund strategies. For example, in 2014–2015, I was very bullish global macro and M&A. I had reasons to think that they would outperform the other strategies, but that didn't happen. I therefore believe that it is important to start with a well-balanced, all-weather portfolio, and then maybe add some minor tilts based on your conviction, but we are talking about small tilts, not strategic allocation changes.

Manager selection is one of the most challenging problems in asset management. How should investors approach evaluating hedge fund managers? Which quantitative and qualitative factors should they consider?

Manager selection is a key topic in hedge fund investing. In selecting and allocating to a manager, you are not buying a track record, you are buying an investment process that you believe will produce an attractive risk-return profile for your portfolio going forward. It is critical to assess the sustainability of this investment process.

You need to understand the strategy and the business. What is the edge of the manager? Does the manager have the capability to continue generating outperformance? We all know about the alpha decay. That means that the manager should have a capability to adapt to a different market environment, to research new strategies and models. Even on the discretionary side, more and more managers use advanced systematic quantitative techniques to screen the universe and support their investment processes. Thus, it is very important for a manager to have the capability to continue researching and improving their investment process.

It is also important to evaluate the viability and sustainability of the manager's business. You need to consider all aspects of the business because there are non-investment risks that can lead to the failure of your investment. The important non-investment risks include operational risk, business risk, and counterparty risk. The counterparty risk arises from directly transacting with a third party which includes the auditors, administrators, brokers, prime brokers, etc. It is also important to evaluate the reputational risk. Then you can look for ways to mitigate the unwanted risks.

The investment risk can be mitigated with customization through managed accounts. The operational risk can be mitigated by working with the manager to improve its infrastructure. The reputational risk can be mitigated by conducting background and reference checks during due diligence process. It is common sense, but there are many areas that you need to consider and the myriad of risks that you need to properly evaluate. It is critical to do that during the due diligence process before the initial allocation, but it is even more critical post-allocation.

You need to keep testing that the assumptions and the understanding that you have regarding the manager, its business, and its investment process are still valid, and the manager is still delivering what you are expecting. The quantitative factors inform and should confirm your qualitative evaluation of the manager.

Over the years you've worked in many successful firms. How diverse were they? In your experience, does diversity help drive better results?

I believe so. I think that cultural and gender diversity brings different perspectives and improves your process. There is no such thing as brainstorming in a group of like-minded people. If everyone thinks the same way, there is no real brainstorming. You end up going with a solution that everyone thinks is the best. Diversity provides an environment that leads to better outcomes for an organization.

When I started my career in 2000, nobody was talking about diversity, but I experienced it from the very beginning of my career. When I was in Hong Kong, we didn't have as much gender diversity in the fixed income trading room. It was male dominated. Unfortunately, I think that is still the case. However, we had a lot of cultural diversity. I worked with people from Hong Kong, China, Korea, Japan, Cambodia, Australia, Europe, and the U.S. I loved working in that environment. At the time, I didn't think of that environment as unusual. It felt natural.

When I joined the Swiss hedge fund, I also worked in a diverse environment. We had people from Switzerland, France, Belgium, Canada, and Eastern Europe. The head of investor relations was from Sri Lanka. We had a woman from Peru who worked in Client Relationship and Compliance. I had someone on my team from Afghanistan. It was great working with people from different cultures.

It was similar when I moved to Canada. Canada is an immigration country, and I know it; but I am still amazed, after 12 years in this country, to hear so many different languages around me when I ride the TTC, Toronto's subway system. I think there are a lot of things that I would not have experienced, at least to the same extent if I had stayed in France. I have been fortunate to work in very diverse environments for the last twenty years. I understand that this is not the case everywhere, but it is difficult for me to imagine what it would be like not to work in a diverse environment.

As you know, this book is about quantitative hedge fund investing. What advice would you give to hedge fund managers and investors that can help them improve their performance?

I struggled with this question. I think it's important to think of an investor–manager relationship as a partnership. When I allocate to managers, I look for long-term partnerships. My advice to the managers is to listen and try to understand the investor's needs, try to address the needs to the extent of your expertise and capabilities. We have a phrase in French "le client est roi," which means "the client is the king." You cannot be successful as a manager if you don't listen to your investors.

If you provide solutions or make suggestions that address the investor's concerns, it will please the investor and help her be more patient when you go through periods of underperformance. It can also potentially lead to new products or strategies that are most likely appealing to other investors. That's a great way for the manager to continue growing his or her business.

My advice to investors is to try to get the most out of the partnership. It is a great opportunity to grow, learn, and improve your own internal processes. It could be the investment process, risk management, execution, or asset allocation. There are excellent managers who have researched asset allocation for decades. There is a lot to learn from and discuss with the managers..

8.3 KEY TAKEAWAYS

Following are the key takeaways from this chapter:

- Historical track records are largely informationless. Interview your manager researchers and map their process to data. "Real performance gains come less from some magical feature you uncover and more from just implementing your current process with more discipline."

- In selecting and allocating to a manager, you are not buying a track record, you are buying an investment process that you believe will produce an attractive risk-return profile for your portfolio going forward.

- Robust hedge fund portfolios are constructed with bottom-up approaches that rely on qualitative and quantitative factors while also controlling for top-down diversification and balance across asset classes.
- There are many types of hedge fund risks that should be carefully evaluated and thoughtfully managed.

CHAPTER 9

Inclusion and Diversity

"I have frequently been questioned, especially by women, of how I could reconcile family life with a scientific career. Well, it has not been easy."—Marie Curie.[a]

"Nobody ... took me seriously. They wondered why in the world I wanted to be a chemist when no women were doing that. The world was not waiting for me."—Gertrude B. Elion.[b]

"Almost always, the creative dedicated minority has made the world better."—Martin Luther King, Jr. [c]

"Women belong in all places where decisions are being made. It shouldn't be that women are the exception."—Ruth Bader Ginsburg.[d]

After discussing many practical aspects of hedge fund portfolio management, our next and last stop on the journey is a discussion of diversity and inclusion, which are important to both Larry and me. This chapter includes personal stories that show why and how diversity and inclusion lead to better investment decisions—a phenomenum supported by empirical research. For example, the 2022 study *Life Cycle Performance of Hedge Fund Managers* by Rose Huang, Elaine

[a]Marie Curie was the first woman to win a Nobel Prize and the only person to win the Nobel Prize in two fields (physics in 1903 and chemistry in 1911).

[b]Gertrude B. Elion was a Nobel Laureate in physiology or medicine, 1988.

[c]Martin Luther King, Jr. was a prominent leader in the civil rights movement known for non-violent resistance.

[d]Ruth Bader Ginsburg was an associate justice of the Supreme Court of the United States.

DOI: 10.1201/9781003175209-9

Jie, and Yue Ma reported that female hedge fund managers had Sharpe ratios that were 17.5 percent above their male peers.[1] This chapter also describes a gender gap in financial academia. We hope that this chapter will inspire young people to pursue their dreams regardless of their background, gender, or race. We also hope that it will motivate seasoned investment professionals to be intentional about promoting diversity and inclusion.

9.1 PROMINENCE OF FEMALE HEDGE FUND MANAGERS WITH MEREDITH JONES

Meredith Jones, a managing director at Ernst and Young, has more than 23 years of investment industry experience, beginning her career at Van Hedge Fund Advisors International, where she was Director of Research. Subsequently she worked in various research capacities in the investment industry at PerTrac Financial Solutions, Barclays Capital, RKCO and Aon. She produced some of the first widely disseminated research on diversity and investing, and built an early Women in Hedge Funds performance index. She continues to specialize in non-financial factors that impact corporate and investment performance, specifically diversity, equity and inclusion, ESG integration, and decarbonization.

Meredith's research has been featured in the international financial media, including *The New York Times, The Wall Street Journal, The Economist, The Journal of Investing, and more.* She is the award-winning author of *Women of The Street: Why Female Money Managers Generate Higher Returns (And How You Can Too)*, was named one of Inc. Magazine's 17 inspiring women to watch in 2017, was selected as a distinguished author by the Securities and Exchange Commission in 2018, and testified before the U.S. House of Representatives in 2019.

Why did you decide to go into finance?

It was an accident. My mom is a retired math professor, and I rebelled against math pretty early in my life. In fact, the only subject I ever made a failing grade in was trigonometry. It happened because I was a typical teenage rebel. My mom was very nerdy. She used to make us convert kilometers to miles on road trips. I didn't know what I wanted to be, but I was pretty sure it was not a mathematician.

When I went to college, I had many different ideas about what I might do, but I realized that genetics really mattered. I inherited some of my mom's affinity for and skills in STEM subjects. My dad was a chemistry professor at some point as well. I had a nerd DNA. After considering many options, I decided to settle on political science and political economy.

After graduation, I joined Vanderbilt to work on research about people making donations to the university through securities. This project involved many aspects of stock trading and valuations. After that I worked at a business magazine where I got to flex my writing muscles. Finally, I saw an ad for an entry level hedge fund analyst. I wasn't entirely sure what a hedge fund was, but the job looked interesting—financial research? Check. Writing market and fund commentary? Check. Just enough math? Check! I applied for that role, got an interview, and hit it off with George Van, the owner of the firm. He ran a hedge fund of funds/family office in Nashville. Interestingly, he believed that only women made good financial researchers. The entire research department consisted of women.

When I joined, I knew a few things about securities, but didn't know anything about hedge funds. I surprised myself by really liking it and seeming to have some talent for it. I worked my way up from an entry level analyst role to be a senior vice president with a seat on the investment committee. I also became the first person to integrate the department by hiring the first man. Although I am a big believer in the financial prowess of women, I also believed that diversification was important. If everyone looks the same, no matter what they look like, that's probably not a good idea. That's how I got my start in the industry.

It seemed normal for you to be on a women-only team. How did you find out that finance is not a women-dominated area?

We had a four-person research department that grew to a team of ten. First, it was all women, then it was all women plus Tim, the first man I hired. So, it felt normal to have a lot of women on an investment team because that was the majority of my experience. We were in Nashville, but I assumed there were a lot of women on Wall Street, until I started going to conferences.

Imagine my surprise at the first conference that I ever went to. As I was walking up to the stage to take my seat as a speaker, there were three guys there already and one of them held up his water glass. I thought he was saying "Cheers" and I waved "Hey." It turns out that he thought I was working for the conference, and he was asking me to refill his glass before we started our panel discussion. To him it was completely opposite, it was not a women-dominated field. We all deal with the biases that we have. It didn't strike me as odd that I was on that panel. I certainly didn't look like the other panelists, but I thought there were a lot of women on other panels. However, to him I looked like an outsider. I filled up his glass because it was the polite thing to do, but then I sat down and kicked butt on the panel. That's when I realized that my experience was different from the norm.

Do you think your career path was typical and straight-forward?

It was and it wasn't. I wasn't an economics major, I didn't get an MBA nor did I go to work at an investment bank, which I think is the traditional path. But, I also think the industry benefits not only from people who look different, but also from people with different backgrounds. The fact that I spent time studying literature and other topics, worked as a writer, and worked in a university environment gave me a skill set that not many of my peers had, and I was able to use that skill set. When I worked at the first hedge fund company, I was writing a congressional testimony for the Long Term Capital Management hearings a year after I was hired. The reason why I was able to do that was precisely because of my non-traditional background. I always tell people that this industry needs many different skill sets. It's not just about crunching the numbers and doing the analysis. If you are a critical thinker, if you are a creative thinker, there is a place for you here. Your background doesn't have to match up to everyone else's. More importantly the skills that you have that you might think are not needed in this industry may be exactly what ends up allowing you to find a place for yourself.

You have remarkable accomplishments, but your journey was probably not always easy. Can you please share some of the obstacles that you had to overcome along the way?

One of the biggest obstacles was that I had a three-year non-compete. Never sign one of those because it derailed me from what I was doing for a long time. Could I have fought it? Probably, but I didn't have deep pockets at the time and it just seemed impossible. Plus, I had signed it and felt I needed to live up to my word.

Another obstacle that I had to overcome was that I was a visible minority—a woman who looked very young. I compounded that my getting braces in my thirties. Suddenly I looked like Mary Katherine

Gallagher from Saturday Night Life going into meetings. I was called a kindergartner a hundred times. When I was doing due diligence on hedge fund managers with billions of dollars in AUM, they would say things like "back in the 1980s when you were in diapers." We all have a tendency to make judgments based on physical characteristics of people. I know I do that. I am not perfect, but I try to be aware of it. That was certainly an impediment because I was not taken seriously. I finally figured out how to turn that into my advantage. I was in a meeting with a couple of partners from overseas, and a manager was dismissive and hadn't answered some questions. One of my partners pulled me aside and told me: "You should've nailed them to the wall. You should've nailed them to the wall. Why didn't you do that!?" I said: "Because I got all the information that I needed. Ultimately, I write the check. I don't need to nail anybody to the wall." If hedge fund managers can't take my questions seriously, that gives me insight into their level of hubris, and hubris is one of the biggest factors that will blow up a hedge fund that I have ever seen. It became a litmus test for me.

I know that you are passionate about mentoring female entrepreneurs. What drives that passion and what are some of the key areas that you tend to focus on as a mentor?

I think that the more women we have in the financial ecosystem, the richer everyone will be. It is not more pie for one group and less pie for another group. The analysis that I have done shows that the more inclusive the global economy is, the better it is for everyone. I find the same pattern whether it is more women in investing who are generating differentiated returns for main street America such as teachers and firemen or whether it is products that appeal to different demographics because consumer purchasing drives the gross domestic product (GDP) or whether it is women making more money because they tend to invest more into their homes and communities. Everything points to the fact that the more women we can bring to the financial industry, the better off we will be as a society. That is an extraordinary compelling argument to me. Game theory says that you should never play a game where you have a winner and a loser. Bringing more women to the financial ecosystem creates a virtuous cycle. For that reason, I find it incredibly appealing.

You can look at that from the smallest things that affect everyday life to the largest things that affect the GDP. Diversified returns, excess returns, job creation, and investing in communities. You can even look at that on a micro level. It has been shown that men who mentor women or advocate for women and diverse individuals at workplaces tend to do better on their own performance evaluations.

I mentor female entrepreneurs who want to take their business to the next level. I am on the board of a financial literacy organization that provides a year of financial literacy to high school girls because the funnel starts to narrow so early. We need to capture that talent and let them know that they can have a fulfilling career in investing. I do that because it benefits me, it benefits you, and it benefits everybody.

When I mentor women, I focus on whatever needs to be done. Everyone needs to get in and roll up their sleeves. It can be anything from being a visible representation in the industry, which I think is incredibly important. If you can't see it, you can't be it. You need to see someone who thinks like you, who dresses like you, and who talks like you. I did a mentoring session for high school girls a few weeks ago. They wanted to know about a career path and a work life balance. If you watch *Billions*, *Succession*, or *The Wolf of Wall Street*, you get this idea that there is no place for people who want to help others or to have kids. Some of my mentoring work is myth busting. It can also get very technical. When I work with entrepreneurs, we set goals at the beginning of a period. We talk about what they want to accomplish from a business perspective whether revenue or client acquisition or public relations. Then we problem solve together to accomplish those goals.

I have many informal mentees. A lot of the time we focus on the next steps in their careers, what they want to accomplish and how to get there. However, I also try to act as a "bitch-and-stitch"—a safe place just to talk about what's going on with their jobs. Women and people who don't have a strong cohort of people who look like them need a safe place where they can say "Damn, this happened today, and it really sucked." It can be very difficult to say those kinds of things because women are stigmatized as being sensitive or emotional. Women also struggle with tooting their own horn. It's nice to be able to call someone to say, "I just totally kicked butt", and have someone who will pump you up.

There are so many avenues for being an advocate and a mentor in this industry. It is important to look for the opportunities, but they don't have to look exactly alike.

I have read about some terrible stats about women struggling to stay in the workplace during the pandemic because of the challenges of juggling work and kids who are learning from home. According to a McKinsey report, one in four women are considering leaving the workforce or downshifting their careers versus one in five men. Do you find yourself encouraging women to stay the course or advising them to make changes?

I don't have children, which makes my life a bit easier. A lot of my friends were home schooling and trying to keep everything together. I think the pandemic has been very difficult for everybody, but it's particularly difficult for women because women still bear a disproportionate amount of support functions whether it is childcare, elder care, cleaning, or cooking.

During the pandemic I have tried to keep people from making decisions that they would regret later. Most of the support I offered was about helping people think through their options. I don't believe people make good decisions when they feel like they don't have options. I tried to help by showing various options, talking through them, and figuring out escape plans. I think that was really important. I had people who did that for me. I hope that I helped some people who were at their low points, because I know that there were people who did that for me.

By the way, we often talk about work life balance as a women problem. I think it's a human problem. One of the reasons we end up with women bearing a disproportionate amount of childcare and elder care is because we still consider it a task for women. Most of the guys I talk to worked too much during the pandemic and they wanted to see their kids' soccer games. They were also worried about losing their jobs. There are a lot of issues that are centered around gender, racial and ethnic diversity, and underrepresented minorities in the industry that deserve special attention, but we can't lose sight that at the end of the day the pandemic affected the human condition for everybody. The best thing you can do is just not be a jerk.

Your award-winning book "The Women of The Street" is fascinating. It shows both empirical evidence and real-life examples of the advantages that women have in risk assessment and portfolio management. What inspired you to write the book?

I entered into the hedge fund industry in 1998, and I didn't meet a woman with a job similar to mine until 2007, which is almost ten years later. It was a very liberating experience to meet a woman with a similar job because I had different behavioral biases than the guys that were around me and I thought that I was weird. It turns out that women have different cognitive, behavioral and background preferences. When I started talking to two female hedge fund managers, a light bulb went off in my head and I started looking for female fund managers. I didn't find many of them for quite some time, but when I did find them, they were almost universally performing well and doing something unique. As I continued looking for them, I finally found enough to be able to do aggregate research on performance in 2010. I spent two years doing research and constructed the first female fund manager index in 2012. It turned out that my intuition was right. Women hedge fund managers were rare, but they were getting it done.

What I really wanted was to make sure that nobody would be in a position like me where they didn't have somebody who could inspire them or could validate their way of thinking. I wanted to give that to people in a way that was a one-to-many transaction, and the book was the best way I could come up with to accomplish that. It's almost like a role model in a box. The women in the book talk about how they got into the career, the challenges they faced, and their strategies. I hoped that the book would attract more women to the industry and keep some who were considering leaving because they felt like outsiders.

I have a high-level question about your research. Why does gender matter in investing?

It goes back to the cognitive and behavioral preferences. It doesn't mean that all men or all women behave the same. It means that there is a preference. I am not trying to pigeonhole people into categories. There is empirical evidence that shows that women trade less, which can be very beneficial for returns. Some of that has to do with confidence. Women tend to be less overconfident in their investments. When you are less overconfident, you don't act every time you think it might be a good time to buy or sell.

There are also stress related responses. Women tend to internalize stress whereas men tend to externalize stress. When a man encounters stress, he wants to solve the problem and fix it. Women generally internalize it. For example, I eat a lot of cookies. What is stress in the financial markets? Market crashes. During the Global Financial Crisis women were less likely to sell because they didn't need to make the pain to go away. They might've been freaking out internally, but they didn't need to take action. Their portfolios performed better during that period because men took actions to make the pain stop and sold at the bottom. It is because of those differences in preferences that women have a differentiated way of approaching investment in general. Of course, there are women who trade like men, and men who trade like women, but that diversification in the investment approach is very positive. We talk a lot about diversification by asset class, geography, and *market cap*. We should also be talking about diversification based on behavior.

We also know from non-financial research that diverse groups make better decisions because they incorporate more perspectives. If you are a venture capital or private equity or real estate fund manager or part of any team that is working on solving a problem, you would benefit from considering more perspectives. It can be uncomfortable. There is research that shows that diversity is uncomfortable, which is why people don't like it. But your outcomes will be better. It really matters to have people in the room who have different experiences and perspectives, not just gender and racial/ethnic diversity. There is even a study that shows that hedge fund managers who grew up poor have differentiated investment patterns. There are a lot of measures of diversity, but people have been largely focusing on the more visible measures of diversity because they are good predictors of cognitive and behavioral alpha.

Although your book is about female investors, I know that you believe in diversity and inclusion. You even testified before the U.S. House of Representatives Financial Services Committee on Diversity and Inclusion in 2019. How did that happen and how did you explain why diversity and inclusion lead to better performance?

I got a random phone call from a Representative's aide asking whether I would testify. I thought it was a joke at first. One of the things I can tell you that it's one of the most stressful things you can do. You don't get a lot of notice to testify. I only had about a week from that initial phone call and was only confirmed the Friday before. You have to get your written testimony in, make all your travel arrangements in a matter of days. Or at least I did. Then once you get there, the security is very tight. That was before January 6th, 2020. Now it's probably even tighter. You only get five minutes, and then they start gaveling you down. It's definitely stressful, but there is a lot of research that shows why diversity and inclusion lead to better performance and I used my testimony to highlight that.

For example, there is a study that shows that women and minorities are disproportionately represented in the top quartile of investment success. If you are looking for a fertile ground for trying to find hedge fund managers, that study would certainly suggest looking at women and minorities. The women in hedge fund index that I created showed a performance differential of about six percentage points per year in six and a half years. It's pretty substantial.

The study on trading behavior showed a one percentage point differential driven by women trading less than men. Morningstar has done research on diverse teams. The National Association of Insurance Commissioners has done research on diverse investments in private equity. There are certainly indications that diversified behavior patterns pay off. Of course, not every time, but if you are searching for alpha, then you want hedge fund managers with behavioral patterns that the market tends to reward. If you look for groups that embrace this diversity, then there are different cognitive and behavioral patterns that are predictive of you generating superior returns.

As you know, this book is about quantitative hedge fund investing. What advice would you give to hedge fund managers and investors that can help them improve their performance?

I believe it matters who manages your money. Don't just look at the number of people on the team, but how included they are. Diversity is so popular today, and it's easy to solve that problem. Anybody can game that—go out and make hires. The inclusion piece is much harder. It's the culture. Can you speak up in a meeting? Do your ideas get heard? It's a much harder hurdle. Don't just look at the front end—the people who are coming in the door. Look at the back end—the people who are leaving the firm. You want to see that the people who are different, who don't look like the founders, are not leaving at a faster rate. You want to see an environment where everyone can get heard because the only way to get the juicy goodness out of diversity is to be inclusive. That has to be on your radar.

9.2 GENDER GAP IN FINANCIAL ACADEMIA WITH MILA GETMANSKY SHERMAN

Mila Getmansky Sherman is the professor and Judith Wilkinson O'Connell Faculty Fellow at the University of Massachusetts Amherst Isenberg School of Management. She is known for her research in empirical asset pricing, systemic risk, hedge funds, financial crises, financial institutions, and system dynamics. Her work has been featured in the highest caliber academic financial journals such as the *Journal of Finance*, *Review of Financial Studies*, and *Journal of Financial Economics*.

Mila, why did you decide to go into finance?

Since my dad was a chemist, I thought I needed to pursue a career in chemistry and an undergraduate degree in chemical engineering from MIT. However, I was fascinated with economics and finance because they seemed so unpredictable. While at school, I worked on a few research projects related to uncertainty and risk. I realized that I needed more education and got accepted to a Ph.D. program at MIT. I was fortunate to meet and learn from very interesting people. For example, I have learned from John Sterman about how to apply system dynamics to better understand systemic risk. He became my advisor, and he was interested in hedge funds. Although I didn't know much about hedge funds, but I found the topic fascinating. Unlike chemistry where there are clearly defined laws, finance cannot be reduced to a set of rules that always hold. Finance is at an intersection of multiple disciplines. I have enjoyed connecting ideas from behavioral sciences, math, machine learning, and even biology.

One of the questions asked at one of my oral exams was about the difference between banking and the other industries. For example, if an automaker goes bankrupt, the other firms in the industry take over the market share. However, if a bank goes bankrupt, the whole financial system might collapse either due to actual interconnectedness or a behavioral "bank run" because depositors are worried about a potential contagion. Finance is not just another field, it's a unique field. I find it fascinating because of its interconnectedness and because it is at an intersection of many areas. I have been able to contribute to this field because I look at issues from the lens of interconnectedness.

You are a well-known academic and a hedge fund expert. Your academic papers have been published in the highest tier journals such as the Journal of Finance, Review of Financial Studies, *and* Journal of Financial Economics. *Can you please share some of the obstacles that you had to overcome along the way?*

One challenge is building a broad knowledge base. I was intentional about taking classes in different disciplines such as system dynamics, optimal control, operations research, math, probabilities theory, physics, finance, economics, and even psychology. Many classes required prerequisites. For example, optimal control required many mechanical engineering classes. Since I didn't have the time to take all the classes, I had to study the prerequisites by myself. Although it was fascinating, it also challenged me to step out of my comfort zone. It was also a bit lonely because I didn't have a cohort of others who followed a similar path and worked on problem sets together like in undergrad. However, it also challenged me to be more social and reach out to people when I needed help.

One of your recent papers Female Representation in the Academic Finance Profession *just came out in the* Journal of Finance.[2] *It talks about the gender imbalance in the academic finance profession, which is related to the broader issue of diversity and inclusion. How did you start thinking about this topic and what are the main conclusions?*

My co-author Heather Tookes and I met a long time ago when we were on the job market after finishing graduate school in 2004. She was at Cornell, and I was at MIT. I met her when I gave a presentation at Cornell, and then we got together again when she came to present her paper at MIT. We got together for dinner and talked about a few potential projects. We ended up writing two papers on convertible bond arbitrage that were published in the *Journal of Financial Economics and the Review of Financial Studies.* Our publications also helped us get tenure.

We enjoyed working with each other and became friends. We meet once in a while, and our children spend time together. We wanted to work on another project together because we enjoyed working together. With all my co-authors, I became friends with them first before choosing to write papers together. When you are friends with people, you enjoy the process even if you don't end up publishing a paper. One day when I was at Heather's house—we were both thinking about the topic of diversity and women in finance. The American Finance Association was already promoting women in finance at conferences. We were both mentoring other women. We were both members of AFFECT (Academic Female Advisory Committee of the American Finance Association). By the way, recently Heather was elected to be the chair of AFFECT. We wanted to serve women in finance. There was a lot of talk about not enough women in finance, not enough publications by women, not enough networks. However, nothing really documented to support or refute those claims. Similar studies had been done in economics and STEM, but nothing in our discipline.

We decided to apply the blueprints from other fields to research what's going on in finance. We didn't plan to publish a paper, we wanted to write a useful report. As we started working on this project, we were gathering data and learning from others about potential research questions. We were not submitting our report to journals. The Journal of Finance editor invited us to publish the paper because he found out about our research, and he wanted the paper to appear in his journal. This is the first paper in this narrow field of diversity in financial academia, but I hope there will be more papers coming out. It is more descriptive. We wanted to understand what was happening. Although the number of faculty members has grown linearly over the 2009–2017 decade, the percentage of women has remained constant at roughly 16 percent. The female representation is lower at top-30 and top-10 at 14.3 percent and 13.1 percent, respectively. The gender gap is even wider for tenured positions. Only 9.7 percent of tenured positions are women.

We have also looked at the number of publications. We have found that generally women have 17 percent fewer publications than men, on average. That is a problem since publications are essential for getting a tenured position in the U.S. We decided to dig deeper to understand why women publish fewer papers. Are they too busy with something else? Are they not capable of writing excellent papers? There could be many potential reasons for that observation. Other researchers have reported that women have more service responsibilities, on average. Therefore, the time constraint is important. For example, during the pandemic women had to spend a disproportionate amount of time to take care of their children and other household responsibilities.

We have decided to look at the quality of publications. We have looked at the journals where people get published and whether papers are co-authored or not. If papers written by women are of lower quality than those written by men, then the share of sole publications in top journals should be lower for women. We found the opposite result. When women get published, the quality of their papers is higher, on average. We have found that the quantity was lower in co-authored papers. Women co-author less frequently, and they tend to co-author with other women. Networks matter! Since the share of women is low, it's much more challenging for them to find other co-authors who are also women. It's like finding a needle in a haystack. Our conclusion is that women need larger networks. One of the goals of AFFECT to help women expand their networks. That can be done by inviting women to conferences and other events.

Interestingly, women have more citations than men. There are many reasons for that. It might be that women write better papers, but it might also be that they could wait longer to publish them due to a higher degree of risk aversion in women. We have also found a gender gap in salaries by looking at public universities where that information is available. Women get paid less. We have also looked at tenure probabilities. If you compare a man and a woman with the same number of publications of the same quality, the woman is 3.1 percentage points less likely to get a tenured position. The gap is high because an additional publication in a top journal improves the probability by approximately 2 percentage points. In other words, the woman must have 1.5 more publications in top journals than the man to get tenured, on average.

The gap used to be even higher, but it has decreased overtime. Unfortunately, the gap in getting full professorship remains very high and it takes women one year longer, on average. It's possible that women take longer because of having kids since most universities give them a one-year leave from the tenure clock. I want to note that it is becoming more common for men to take a one-year paternity leave.

It is encouraging that our research is showing that overall gender gap is declining in the U.S. In case of tenure, it is strong up to six years but then disappears after eight years. However, when I talked to my colleagues in Europe about this research, they told me that the gap was higher in Europe and the situation was not getting any better. We only looked at the top 100 business schools in the U.S. It is hard to objectively compare the gender gap in the U.S and Europe because they have a different set of requirements for tenure. For example, publications are less important there. Unfortunately, the full professorship gap is still persistent. It is also about 3.1 percentage points, roughly equivalent to 1.5 publications in top journals.

What can be done to close the gender gap? How do you help women?

As I have mentioned earlier, it is important to help women expand their networks. AFFECT facilitates networking among women. My advisor Andrew Lo was always great at inviting me and his other PhD students to present at conferences. I really benefited from having Andrew as a mentor, and I enjoy mentoring other women, particularly women in finance. One of the reasons why women represent only 16 percent of faculty is the pipeline—we have fewer women graduating with finance degrees.

We started the Women in Finance organization for my undergraduate students to help them get comfortable with a career in finance. We applied and got accredited to start a chapter of Smart Woman Securities, a national non-profit organization with a mission to educate undergraduate women on investment and finance, at the University of

Massachusetts Amherst. We meet once a week. We go through a specific syllabus together. However, networking is a key part of this initiative. We go to national meetings. We organize trips to Boston and New York City for presentations. We organize mentorship opportunities. We also try to be inclusive—our chapter is open to men and to students from all five schools of the university. We find that it is very important for women to get comfortable with the idea of becoming finance professionals.

It all started with Coleen, a student in honors financial modeling class. She looked at me one day and told me: "I am the only woman in finance here. What can we do about that?" I told her that I had been waiting for that conversation for a long time, but it needed to be student driven. She started the chapter, and it grew to seventy people. Coleen graduated, got a nice finance job in Boston. She is the one who paved the way for many women at the school. We are even talking about starting an ESG fund because women tend to be passionate about ESG. We are in the process of raising money for the fund.

Why are you so passionate about helping women?

My sister has accomplished a lot. She is a CEO of a biotech company. I have been fortunate to have parents who told us that we could do anything, but also wanted us to remember our roots, stay humble and give back. They taught me that I should always look around and help people around me. I see a lot of girls who tell themselves that they can't be good at math or finance. I want to help them to believe in themselves and find opportunities. Finance is always changing, and it is full of opportunities. We have talked about ESG. Crypto and *Non-fungible tokens* (NFTs) are not even in finance books yet, but people are trading them. There are many opportunities for women to find an area where they feel that they belong.

Do you think diversity is important?

Portfolio diversification is widely accepted. It is not prudent to put all your eggs in one basket. I think that the diversity of opinion is similar to portfolio diversification. As you add more assets to your portfolio, the overall risk goes down. Investors may want to diversify across portfolio managers or teams of portfolio managers. My hypothesis is that the diversity of opinion should matter at least in terms of minimizing risk. Risk reduction improves the longevity of investments, which should have a positive impact on total returns.

Do you think there is a significant gap between academia and the industry? What can industry professionals and academics learn from each other?

I think it is very important for academics to learn from industry professionals. Most of my research is driven by the industry. When Heather Tookes and I decided to work on convertible arbitrage research, we had a hypothesis about arbitrageurs and how they affect the market, its liquidity and efficiency. We started by scheduling meetings with hedge fund managers who specialized in convertible bond arbitrage. The conversations were extremely helpful because we made sure that our hypotheses were reasonable, and our research was useful for hedge fund managers. I followed a similar approach with all my hedge fund research. I wanted to make sure that I was not building toy models that were relevant for a brief period.

The University of Massachusetts Amherst has the Center for International Securities and Derivatives Markets (CISDM). Every year we organize conferences that have speakers from both academia and industry. Recently we had panels on ESG, crypto, technology, and the future of finance. The majority of attendees are industry professionals who want to learn from academic research, but academics also try to learn from the industry.

I teach a course on alternative investments. I invite practitioners as guest speakers because I think students can learn a lot from them. I have personally learned a great deal from the industry, and I encourage all academics to learn from the industry as well. Sometimes it pushes us out of our comfort zone. When I was a Ph.D. student at MIT, Andrew Lo told me to work at a bank. I worked in the Asset Management Division at Deutsche Bank. It helped me appreciate work of practitioners and understand the pressure and constraints that they face. I have also learned a lot about how decisions are made. They are made by people, not robots. We can't learn everything from books. It is important to talk to people and learn from them.

In our book, we reference several of your papers on hedge funds. Are there any interesting research topics you are currently exploring? What advice would you give to hedge fund managers and investors that can help them improve their performance?

I recently wrote the paper *Global Realignment in Financial Market Dynamics* with Monica Billio, Andrew Lo, Loriana Pelizzon, and Abalfazl Zereei.[3] We presented this paper at the American Finance

Association conference. We used minute-level data of country Exchanged Traded Funds (ETFs) from 2012 to 2020 and investigated the centrality and connectedness of various countries. We found that the U.S. was central to the global financial system before 2018, but the U.S.-China trade war and the pandemic had partially shifted the centrality from the U.S. to China. We moved from a unipolar to a bipolar world.

The implications for investors are huge. In the past, investors needed to think about what was happening in the U.S. and consider implications for the rest of the world. Now investors must think how the rest of the world affects the U.S. The global financial system is much more interrelated. I see articles about in New York Times and Economist about geopolitical links, but it is fascinating that the financial markets are already incorporating that interconnectedness.

Since global portfolios are still largely U.S.-centric, there are practical implications of this important change. Investors can no longer ignore potential shocks in foreign countries because they might have a more pronounced impact on the U.S. markets than observed in the past. The systemic risk and the contagion effect are more significant. It is also interesting that indirect effects are becoming more important. For example, if India has a low direct effect on the U.S., it might still have a high indirect effect on the U.S. if India has a strong direct effect on China and China has a strong direct effect on the U.S. Thus, it is important to not over-rely on historical cross-correlations and volatilities.

9.3 KEY TAKEAWAYS

Following are the key takeaways from this chapter:

- Diversity and inclusion improve investment performance. "The only way to get the juicy goodness out of diversity is to be inclusive."
- Cognitive and behavioral preferences of women lead them to enter fewer unprofitable trades and to not exit positions prematurely due to stress.
- Gender gap in financial academia is large. Only 9.7 percent of tenured positions are women. U.S. business schools require women to have 1.5 more publications in top academic journals relative to men to earn full professorships.
- Women write higher quality papers and have more citations.

Conclusion

"I will not propose to you that my way is best. The decision is up to you. If you find some point which may be suitable to you, then you can carry out experiments for yourself. If you find that it is of no use, then you can discard it."—Gyalwa Rinproche, the 14th Dalai Lama.[a]

"Live as you were to die tomorrow. Learn as if you were to live forever."—Mahatma Gandhi.[b]

We hope that you have enjoyed the journey through the fascinating world of hedge fund investing. The hedge fund industry has a long and rich history of innovation. Its strategies have evolved from the early long-short technical signals of Alfred Winslow Jones in 1949 to modern systematic trading and machine learning approaches. Hedge fund investing is often misunderstood. As promised, we have debunked several myths: hedge fund investing and selection of top performers are easy; hedge funds hedge; active and socially responsible hedge funds outperform; and investors benefit from hedge funds identifying undervalued stocks.

Empirical hedge fund research is performed by academics and practitioners. While each perspective is valuable, they are also incomplete—making it vital to appreciate and blend both views in order to build robust hedge fund portfolios. We showed that hedge fund researchers can use one or several databases. However, the quality of the databases varies significantly and empirical data should be carefully evaluated and adjusted for biases. In addition, investors should either allocate via managed accounts or carefully account for rebalancing frictions in their investment process.

[a]Gyalwa Rinproche is the 14th Dalai Lama, the highest spiritual leader of Tibet.
[b]Mahatma Gandhi was one of the greatest political and spiritual leaders of the 20th century known for pioneering non-violent civil disobedience.

After a deep-dive into the empirical data, we turned to the fascinating and challenging topic of hedge fund selection. We demonstrated that fund selection should be customized to the specific objectives and constraints of each unique investor. We also showed the importance of identifying hedge funds with a positive marginal impact on the existing portfolios. We then discussed how machine learning offers several promising approaches that can help with selecting a subset of factors that are relevant for a given hedge fund. However, quantitative performance evaluation is challenging because of the role of serendipity. Therefore, a hedge fund investor should also consider qualitative factors. We also highlighted the need to perform specialized operational due diligence of digital asset funds.

When we discussed portfolio construction, we also showed the importance of customizing portfolios. For example, pension plan investors can benefit from explicitly evaluating the impact of candidate investments on their funding ratios. Although most portfolio techniques perform poorly out-of-sample, hedge fund investors can improve performance by diversifying risk across strategies and targeting volatility. Investors may also consider top-down portfolio approaches or recent cutting-edge techniques covered in the book. However, it is important to not be overreliant on recent historical data. When the market environment changes, historical periods with a similar environment, rather than the most recent period, are more relevant for forward-looking portfolio decisions.

Although we learned a lot from the challenging process of distilling hundreds of heavy academic papers into pragmatic toolboxes of research-based investment ideas and decisions, we particularly enjoyed gaining insights from the thought leaders in modern finance who generously contributed to our book.

Katy Kaminski taught us that trend following is about applying mathematical techniques to measure where markets are going and what the markets are doing. Trend following has stood the test of time, and it has provided crisis alpha.

Kai Wu showed that machine learning offers a powerful toolbox of methods but "ultimately, you have to pick the right tool for the job." Unstructured data offers many opportunities for creative applications of machine learning to trading.

Asha Mehta discussed how access to capital, technology, and stronger governance and development programs have turned emerging markets not only into a compelling investment opportunity for investors but also a fundamentally different backdrop in the global landscape.

Adam Duncan showed that historical track records are largely informationless, highlighting the importance of interviewing in-house manager researchers and mapping their process to data. "Real performance gains come less from some magical feature you uncover and more from just implementing your current process with more discipline."

Christophe L'Ahelec taught us that in selecting and allocating to a manager, we are not buying a track record. Instead, we are buying an investment process that we believe will produce an attractive risk-return profile for your portfolio going forward. Robust hedge fund portfolios are constructed with bottom-up approaches that rely on qualitative and quantitative factors while also controlling for top-down diversification and balance across asset classes. There are many types of hedge fund risks that should be carefully evaluated and thoughtfully managed.

Meredith Jones demonstrated that diversity and inclusion improve investment performance. "The only way to get the juicy goodness out of diversity is to be inclusive." Cognitive and behavioral preferences of women lead them to enter fewer unprofitable trades and to not exit positions prematurely due to stress.

Mila Getmansky Sherman showed that the gender gap in financial academia is large. Only 9.7 percent of tenured positions are women and U.S. business schools require women to have 1.5 more publications in top academic journals relative to men to earn full professorships—yet women write higher quality papers and have more citations.

We hope that this book has given you the necessary tools for building a hedge fund portfolio for investors by following rigorous steps of determining which building blocks you can rely on and thoughtfully putting them together. Feel free to reach out to Larry (lswedroe@buckinghamgroup.com) or Marat (marat@efficient.com) with your questions and feedback.

APPENDIX A

Manager Selection and Hedge Fund Factors

A.1 FRAMEWORK OF MOLYBOGA, BILSON, AND BAEK

This section describes performance ranking of hedge funds from the 2017 paper *Assessing Hedge Fund Performance with Institutional Constraints: Evidence from CTA Funds* by Molyboga, Baek, and Bilson.[1]

For each fund i its t-statistic of alpha with respect to the CTA benchmark was calculated based on the lagging 60 months of returns. The calculation was performed at time t, such as the end of December 1998 for the first investment decision.

First, an OLS regression was utilized as follows:

$$r^i_\tau = \alpha^i_t + \beta^i_t I_\tau + \epsilon^i_\tau, \tag{A.1}$$

where $\tau = t-60, t-59, \ldots, t-1$, r^i_τ was the net-of-fee excess return of fund i at time τ, and I_τ was the excess return of the CTA benchmark.

Part of the estimation procedure included estimating the standard error of alpha, $\sigma(\alpha^i_t)$, which was then used to calculate the t-statistic of alpha for fund i at time t as

$$T^i_t = \alpha^i_t / \sigma(\alpha^i_t), \tag{A.2}$$

which was then used to rank the eligible funds.

DOI: 10.1201/9781003175209-A

A.2 FACTOR SELECTION

This section describes factor selection.

Consider a time-series $y = (y_1, ..., y_T)$. A linear regression approach expresses y as a linear combination of N factors $f^1, ..., f^N$ with each factor f^i represented by its time-series $(f_1^i, ..., f_T^i)$:

$$y_t = \alpha + \sum_{i=1}^{N} \beta^i f_t^i + \epsilon_t, \tag{A.3}$$

α is the intercept, $\beta = (\beta^1, ..., \beta^N)$ is the vector of slope coefficients and ϵ_t is the residual at time t.

The typical Ordinary Least Squares (OLS) minimizes the squared error:

$$(\hat{\alpha}^{OLS}, \hat{\beta}^{OLS}) = \arg\min \sum_{t=1}^{T} \left(y_t - \alpha - \sum_{i=1}^{N} \beta^i f_t^i \right)^2 . \tag{A.4}$$

The LASSO approach, introduced in the 1996 paper *Regression Shrinkage and Selection via the Lasso* by Robert Tibshirani, also minimizes the squared error but includes an L_1-norm penalty on factor loadings:[2]

$$(\hat{\alpha}^{LASSO}, \hat{\beta}^{LASSO}) = \arg\min \sum_{t=1}^{T} \left(y_t - \alpha - \sum_{i=1}^{N} \beta^i f_t^i \right)^2 + \lambda \sum_{i=1}^{N} |\beta^i|. \tag{A.5}$$

The 2005 paper *Regularization and Variable Selection via the Elastic Net* by Hui Zou and Trevor Hastie showed that when potential predictors are correlated, the Elastic Net approach that combines L_1-norm (LASSO) and L_2-norm (ridge) penalties results in superior prediction accuracy.[3] The Elastic Net approach is defined as follows:

$$(\hat{\alpha}^{EN}, \hat{\beta}^{EN}) = \arg\min \sum_{t=1}^{T} \left(y_t - \alpha - \sum_{i=1}^{N} \beta^i f_t^i \right)^2 + \lambda_1 \sum_{i=1}^{N} |\beta^i| + \lambda_2 \sum_{i=1}^{N} (\beta^i)^2. \tag{A.6}$$

The 2020 study *Empirical Asset Pricing via Machine Learning* explained that LASSO is effective at factor selection but Elastic Net also mitigates the issue of estimated coefficients being too large. [4]

APPENDIX B

Performance Persistence

B.1 BOOTSTRAPPING

This section includes the implementation details of the bootstrapping approach used in the 2006 study *Can Mutual Fund "Stars" Really Pick Stocks? New Evidence from a Bootstrap Analysis*, the 2007 study *Do Hedge Funds Deliver Alpha? A Bayesian and Bootstrap Analysis*, and the 2010 paper *Luck versus Skill in the Cross-Section of Mutual Fund Returns.*[1,2,3]

Bootstrapping is performed using the following steps:

1. The hedge fund data is adjusted for biases, as discussed in detail in Section 2.2.1. The pool of hedge funds is selected using investment restrictions specific to institutional investors. For example, in their study *Do Hedge Funds Deliver Alpha? A Bayesian and Bootstrap Analysis*, Kosowski, Naik, and Teo imposed minimum track record length and AUM requirements.[4] This step produces a bias-free dataset of N time-series of hedge fund returns for period $t = 1, ..., T$. The dataset has a lot of null values because track records of defunct funds end before time T and active hedge funds may have track records that start after $t = 1$.

2. A factor model is selected given the set of hedge fund managers. This topic is discussed comprehensively in section 3.2.

3. For each fund from the pool, its alpha and the t-statistic of alpha are estimated. Kosowski, Naik, and Teo recommended

DOI: 10.1201/9781003175209-B

applying the Newey-West adjustment for serial correlation and heteroskedasticity. The distribution of the t-statistics of alpha is referred to as the actual distribution. The aforementioned 2006 study *Can Mutual Fund "Stars" Really Pick Stocks? New Evidence from a Bootstrap Analysis* by Robert Kosowski, Allan Timmermann, Russ Wermers, and Hal White recommended using the t-statistics of alpha rather than alphas to measure performance due to the superior statistical properties since precision of alpha estimates vary with the length of funds' track records and funds' volatilities.[5]

4. A zero alpha dataset is constructed by subtracting each fund's estimate of alpha from its monthly returns. Thus, the zero alpha dataset preserves the properties of fund returns while having the true alpha of zero. The zero alpha dataset also covers the period $t = 1, ..., T$. An alternative approach to constructing a zero alpha dataset is to consider regression residuals since they have a zero alpha, by construction, and also preserve the return characteristics.

5. A single simulation run is a random sample with replacement from the zero alpha dataset. There are three main approaches to random sampling that vary along two dimensions (i.e., sampling by fund or by joint cross-section of funds; by a single month or a block of months):

   - Kosowski, Timmermann, Wermers, and White performed sampling fund by fund, implicitly assuming zero cross-sectional correlation, and selected single month's returns, implicitly assuming zero serial correlation.[6]
   - Fama and French performed sampling by cross-section, which attempts to retain the cross-correlation structure, and selected single month's returns, also assuming zero serial correlation. They compared their results to those of Kosowski, Timmermann, Wermers, and White and argued that failing to use joint sampling likely understated the cross-sectional deviation of the t-statistics of alpha under the null hypothesis and, therefore, overstated the evidence for positive and negative skill.[7]
   - Finally, Kosowski, Naik, and Teo sampled blocks τ of the cross-section in an attempt to preserve the cross-correlation

and serial-correlation structure, which is important when considering hedge funds.[8] While there are statistical approaches to selecting an optimal value of τ, a simple solution includes repeating the analysis using several values of τ such as $\tau = 1, 3, 6, 12$. For example, if February 2020 is randomly selected and τ is equal to one, then a slice of returns for all funds on February 2020 are recorded in a simulation before randomly selecting another date. If τ is equal to three, then a block of returns for all funds on December 2019, January 2020, and February 2020 are recorded in a simulation before randomly selecting another date. A larger value of τ assumes longer serial dependencies.

The simulation run produces a set of N time-series of length T. Time-series with an insufficient number of non-null returns are excluded. The t-statistics of alpha are calculated for each remaining time-series and saved.

6. Simulations are performed a large number of times such as 10,000. The distributions of the t-statistics of alpha values aggregated across all simulations form a simulated distribution, which represents the distribution of the t-statistics of alpha under the null hypothesis of zero alpha. The actual distribution is compared to the simulated distribution to detect evidence of positive skill and negative skill by examining the right and the left tails, respectively. For example, if the simulated distribution shows 50 out of 10,000 funds having the t-statistic of alpha that exceeds 2 and the actual distribution has 100 out of 10,000 funds, that is interpreted as evidence of skill, or some funds with true positive alpha.

B.2 FALSE DISCOVERY

This section includes the implementation details of the false discovery approach from the 2010 paper *False Discoveries in Mutual Fund Performance: Measuring Luck in Estimated Alphas* by Laurent Barras, Olivier Scaillet, and Russ Wermers.[9]

It is performed using the following steps:

1. **Estimate the proportion of zero alpha funds** π_0. This step leverages the fact that the majority of p-values above a sufficiently

large threshold p_L such as 0.6, or t-statistics of alpha that are small, come from zero-alpha funds. Since the distribution of p-values of zero-alpha funds is uniform, the share of funds with p-values greater than p_L should be equal to $\pi_0(1 - p_L)$. Therefore, if the total number of hedge funds is equal to N and the number of funds with p-values greater than p_L is equal to $\hat{W}(p_L)$, the estimated proportion of zero-alpha funds is equal to

$$\hat{\pi}_0(p_L) = \frac{\hat{W}(p_L)}{N} \frac{1}{1 - p_L}. \tag{B.1}$$

The value of p_L can be estimated using the bootstrap approach of the 2002 paper *A Direct Approach to False Discovery Rates.* [10] However, Barras, Scaillet, and Wermers reported that the estimate of $\pi_0(p_L)$ in expression (B.1) was not sensitive to the choice of p_L and recommended using any value between 0.5 and 0.6.[11] It is worth noting that the assumption of only zero-alpha funds having p-values above a sufficiently large threshold may be too strong because of a very low signal-to-noise ratio in fund returns. For example, a fund with a true alpha of 2 percent may produce an alpha of 0 due to back luck and, thus, be included in the estimate of zero-alpha funds. This issue was discussed in detail in the 2019 study *Reassessing False Discoveries in Mutual Fund Performance: Skill, Luck, or Lack of Power?*[12] The authors reported that the false discovery rate methodology was too conservative and underestimated the proportion of nonzero-alpha funds.

2. **Estimate the proportion of false discoveries in the right and the left tail for a given significance level γ.** Since the proportion of zero-alpha funds is equal to π_0, the estimate of zero-alpha funds in the right tail $\hat{F}_\gamma^+$ and the left tail $\hat{F}_\gamma^-$ can be calculated as

$$\hat{F}_\gamma^+ = \hat{F}_\gamma^- = \hat{\pi}_0 \frac{\gamma}{2}. \tag{B.2}$$

3. **Estimate the proportion of funds with positive and negative skill at significance level γ.** The proportion of funds in the right tail (t-statistics of alpha greater than the threshold level that corresponds to γ)

$$E[S_\gamma^+] = E[T_\gamma^+] + E[F_\gamma^+] \tag{B.3}$$

consists of the positive-alpha funds with sufficiently high t-statistic of alpha and the zero-alpha fund who happened to be lucky. Note that some positive-alpha funds have a t-statistic below the threshold because of poor luck and negative-alpha funds are not expected to have a large enough t-statistic of alpha to be included.

Similarly, the proportion of funds in the left tail at significance level γ is equal to

$$E[S_\gamma^-] = E[T_\gamma^-] + E[F_\gamma^-] \tag{B.4}$$

consists of the negative-alpha funds with sufficiently negative t-statistic of alpha and the zero-alpha fund who happened to be unlucky. Some negative-alpha funds have a t-statistic above the threshold because of luck and positive-alpha funds are not expected to have a sufficiently negative t-statistic of alpha to be included.

Given the estimate of $\hat{F}_\gamma^+$ and $\hat{F}_\gamma^-$ from expression (B.2), the proportion of positive-alpha funds can be estimated as

$$\hat{T}_\gamma^+ = \hat{S}_\gamma^+ - \hat{\pi}_0 \frac{\gamma}{2}, \tag{B.5}$$

where $\hat{S}_\gamma^+$ can be estimated from the empirical distribution of the t-statistics of alpha.

Similarly, the proportion of negative-alpha funds can be estimated as

$$\hat{T}_\gamma^- = \hat{S}_\gamma^- - \hat{\pi}_0 \frac{\gamma}{2}, \tag{B.6}$$

where $\hat{S}_\gamma^-$ can be estimated from the empirical distribution of the t-statistics of alpha.

Finally, the proportion of positive-alpha and negative-alpha funds in the population is estimated as

$$\hat{\pi}^+ = \hat{T}_{\gamma_L}^+ \tag{B.7}$$

and

$$\hat{\pi}^- = \hat{T}_{\gamma_L}^-, \tag{B.8}$$

respectively. While the authors provided a bootstrapping approach for estimating the value of γ_L, they report that any value between 0.35 and 0.45 worked sufficiently well.[13] Note that these estimates implicitly assume that all observed t-statistics of alpha of positive-alpha funds are in the right tail of the distribution and all observed t-statistics of alpha of negative-alpha funds are in the left tail of the distribution, which is unlikely because of a low signal-to-noise ratio in fund return. This issue is also discussed in the aforementioned 2019 paper.[14]

4. **Estimate the locations of positive and negative skill funds.** Barras, Scaillet, and Wermers provided the estimation details in their internet appendix.[15] The authors used the fact that for positive and negative skill funds, their t-statistics followed a non-central student distribution with $T-5$ degrees of freedom and a non-centrality parameter of $\sqrt{T}\alpha/\sigma_\epsilon$, where α is equal to $\alpha^+ > 0$, the true alpha for positive skill funds, or $\alpha^- < 0$, the true alpha for negative skill funds. T is the length of the track record and σ_ϵ is the standard deviation of residuals when the fund's returns are regressed against the factors. Therefore,

$$E[T_\gamma^+] = \pi^+ P[t > t_{T-5,1-\gamma/2}|\alpha^+] \tag{B.9}$$

and

$$E[T_\gamma^-] = \pi^- P[t < t_{T-5,\gamma/2}|\alpha^-] \tag{B.10}$$

with $t_{T-5,\gamma/2}$ and $t_{T-5,1-\gamma/2}$ denoting the quantiles of the noncentral student distribution.

Estimates $\hat{\pi}^+$ and $\hat{\pi}^-$ are produced in the previous step using expressions (B.7) and (B.8). The authors suggest using $\gamma = 0.2$ and expressions (B.5) and (B.6) to estimate $\hat{T}_\gamma^+$ and $\hat{T}_\gamma^-$ and then rely on equations (B.9) and (B.10) to produces estimates $\hat{\alpha}^+$ and $\hat{\alpha}^-$. The locations of the t-statistics of alpha for positive alpha and negative alpha funds are then calculated as $\hat{t}^+ = \sqrt{T}\hat{\alpha}^+/\hat{\sigma}_\epsilon$ and $\hat{t}^- = \sqrt{T}\hat{\alpha}^-/\hat{\sigma}_\epsilon$, respectively. The authors modify the estimation procedure to account for cross-sectional and serial dependence, but their estimates are similar to the base case and remain unbiased.

B.3 PERFORMANCE EVALUATION WITH NOISE REDUCED ALPHA

This section covers the implementation details of the noise reduced alpha approach from the 2018 study *Detecting Repeatable Performance* by Campbell Harvey and Yan Liu.[16]

The authors followed a random effects methodology that assumed that each fund's alpha was driven independently from a common cross-sectional distribution, which was characterized as a Gaussian Mixture Distribution (GMD), a weighted sum of Gaussian (normal) distributions. For a GMD with L components, its model parameter set $\theta^L = \{(\mu_l)_{l=1}^L, (\sigma_l^2)_{l=1}^N, (\pi_l)_{l=1}^L\}$, where the means are specified in an ascending order $\mu_1 < ... < \mu_L$, σ_l^2 are the variances, and the non-negative weighs add up to 1 as $\pi_1 + ... + \pi_L = 1$. If we consider positive-alpha, zero-alpha and negative-alpha groups of funds, then L is equal to 3. However, unlike the false discovery rate approach previously discussed, the true alpha is a continuous variable and its density is estimated for each fund and for the cross-section of funds. We denote by R_i the time-series of returns of fund i, where $i = 1, ..., N$.

The maximum likelihood estimation procedure relied on the expectation maximization (EM) approach, which treated alphas as missing observations and went through iterations of expectation and maximization steps to update the alpha distribution and the model parameters:

- **Expectation step.** Given the model parameters that include (i) fund-specific factor loadings denoted by B, (ii) residual standard deviation denoted by Σ, and (iii) the GMD parameters discussed above denoted by θ, alphas were randomly drawn from the conditional distribution of alphas, and the likelihood function was averaged across the random draws to learn about the distribution of manager skill.

- **Maximization step.** Given the distribution of the manager skill (alpha), the model parameters were updated.

The estimation procedure includes the following steps:

1. **The initial step**. The initial parameters $G_0 = [B_0, \Sigma_0, \theta_0]$, where the loadings B_0 and the covariance matrix of the residuals Σ_0 are estimated using fund-by-fund OLS and θ_0 is estimated using OLS based on the estimated OLS alphas.

2. **The expectation stage of step** k. Given the estimated parameters $G_k = [B_k, \Sigma_k, \theta_k]$, the log-likelihood function is estimated following the principles of Monte Carlo EM, which suggests replacing expectation with a sample mean obtained from

M simulations as follows:

$$\hat{L}(G|G_k) = \frac{1}{M}\sum_{m=1}^{M}\sum_{i=1}^{N}\log\left[f(R_i|\alpha_i^m,\beta_i,\sigma_i)f(\alpha_i^m|\theta)\right] = \frac{1}{M}\sum_{m=1}^{M}\sum_{i=1}^{N}\log f(R_i|\alpha_i^m,\beta_i,\sigma_i) + \frac{1}{M}\sum_{m=1}^{M}\sum_{i=1}^{N}\log f(\alpha_i^m|\theta). \quad \text{(B.11)}$$

The authors reported that $M = 100$ was sufficient as it produced estimates that were close to those generated using $M = 1,000$.

3. **The maximization stage of step** k. Given the estimate $\hat{L}(G|G_k)$, the parameters that maximize the expected log-likelihood are used to estimate G_{k+1}. The decomposition in expression (B.11) allows for estimating B_{k+1} and Σ_{k+1} by maximizing the first term of the log-likelihood function and estimating θ_{k+1} by maximizing the second term of the log-likelihood function.

4. **Step** $k+1$. Given the new set of parameters G_{k+1}, steps 2 and 3 are repeated until the convergence of the parameter estimates.

The authors provided the implementation details for steps 2 and 3 of the algorithm in the appendix. As part of the maximization stage in step 3, the methodology produced estimates of individual fund alphas by pooling information from the cross-sectional distribution.

B.4 PERFORMANCE EVALUATION WITH SEEMINGLY UNRELATED ASSETS

This section includes technical details and implementation steps of the seemingly unrelated assets approach from the 2002 paper *Mutual Fund Performance and Seemingly Unrelated Assets* by Lubos Pastor and Robert Stambaugh.[17]

Let us consider N time-series of hedge fund returns with r_t^i representing an excess return of hedge fund i at time t and K factors $f_t^1, \ldots, f_t^K$. The standard regression approach:

$$r_t^i = \alpha^i + \sum_{k=1}^{K}\beta_k^i f_t^k + \epsilon_t^i, \quad \text{(B.12)}$$

where α^i is the intercept, which represents abnormal performance of the hedge fund i, $\beta^i = (\beta_1^i, ..., \beta_K^i)'$ is the vector of slope coefficients and ϵ_t^i is the residual at time t.

Hedge funds are typically evaluated based on the t-statistic of alpha, which is often measured using a Newey-West adjustment for heteroskedasticity and serial correlation. As previously discussed, this adjustment has no impact on the estimates of α and β^i but it influences the statistical inference (i.e., adjusts values of the t-statistics of alpha and betas).

Kosowski, Naik, and Teo used hedge fund benchmarks $F_t^1, ..., F_t^M$ as seemingly unrelated assets, and examined two set of regressions.

The first set of regressions involved regressing the performance of the M hedge fund benchmarks on the K factors $f_t^1, ..., f_t^K$:

$$F_t^m = \alpha_F^m + \sum_{k=1}^{K} \beta_{F,k}^m f_t^k + \epsilon_{F,t}^m. \tag{B.13}$$

The second set of regressions was similar to (B.12), except hedge fund benchmarks were added to the factors on the list of explanatory variables as follows:

$$r_t^i = \delta^i + \sum_{m=1}^{M} \gamma_{F,m}^i F_t^m + \sum_{k=1}^{K} \gamma_{f,k}^i f_t^k + u_t^i. \tag{B.14}$$

Using the expression (B.13) to substitute F_t^m in the equation (B.14) resulted in

$$r_t^i = \left[\delta^i + \sum_{m=1}^{M} \gamma_{F,m}^i \alpha_F^m\right] + \sum_{k=1}^{K} \left[\gamma_{f,k}^i + \sum_{m=1}^{M} \gamma_{F,m}^i \beta_{F,k}^m\right] f_t^k + \left[u_t^i + \sum_{m=1}^{M} \gamma_{F,m}^i \epsilon_{F,t}^m\right]. \tag{B.15}$$

Since f_t was uncorrelated with both $\epsilon_{F,t}^m$ and u_t^i,

$$\alpha^i = \delta^i + \sum_{m=1}^{M} \gamma_{F,m}^i \alpha_F^m \tag{B.16}$$

and

$$\beta_k^i = \gamma_{f,k}^i + \sum_{m=1}^{M} \gamma_{F,m}^i \beta_{F,n}^m. \tag{B.17}$$

The decomposition used in the Pastor-Stambaugh approach resulted in more precise estimates of alphas because it employed an empirical Bayesian approach to draw additional information from longer track records of hedge fund benchmarks.

The prior distribution of Σ, the covariance matrix of the residuals $\epsilon_{F,t}^m$ in equation (B.13) followed an inverted Wishart distribution:

$$\Sigma^{-1} \sim W(H^{-1}, \nu). \tag{B.18}$$

The degrees of freedom $\nu = m + 3$ implied that the prior had little information about Σ. Since the prior expectation of Σ was equal to $H/(\nu - m - 1)$, the authors specified $H = s^2(\nu - m - 1)I_m$ with I_m denoting an m-by-m identity matrix, so that $E[\Sigma] = s^2 I_m$.

The value of s^2 was set to the average of the diagonal elements of the OLS estimate of Σ from equation (B.13), consistent with an empirical Bayes approach.

The prior for $\alpha_F = (\alpha_F^1, ..., \alpha_F^M)'$ was specified as a normal distribution:

$$\alpha_F|\Sigma \sim N\left(0, \sigma_{\alpha_F}^2 \left[\frac{\Sigma}{s^2}\right]\right) \tag{B.19}$$

with σ_{α_F} serving as the skill uncertainty. Pastor and Stambaugh called it a "mispricing" uncertainty, which was more appropriate for traditional passive benchmarks. Setting $\sigma_{\alpha_F} = 0$ represents perfect confidence in $\alpha_F = 0$ (i.e., the hedge fund benchmarks have zero alpha or skill relative to the factors). Setting $\sigma_{\alpha_F} = \infty$ represents diffuse prior. Setting a non-zero finite value of σ_{α_F} allowed for a degree of skill exhibited by the hedge fund benchmarks with respect to the factors.

Denote $\gamma^i = (\gamma_{F,1}^i, ..., \gamma_{F,M}^i, \gamma_{f,1}^i, ..., \gamma_{f,K}^i)'$. The prior beliefs regarding the parameters in the regression (B.14) were specified as follows:

- The prior for $\sigma_{u^i}^2$, the variance of the error terms u_t^i, followed an inverted gamma distribution:

$$\sigma_u^2 \sim \frac{\nu_0 s_0^2}{\chi_{\nu_0}^2}, \tag{B.20}$$

where $\chi_{\nu_0}^2$ was the chi-square distribution with ν_0 degrees of freedom.

- The priors for δ and γ^i, conditional on σ_u^2, were normal, independent of each other:

$$\delta|\sigma_u^2 \sim N\left(\delta_0, \left(\frac{\sigma_u^2}{E[\sigma_u^2]}\right)\sigma_\delta^2\right) \tag{B.21}$$

and

$$\gamma|\sigma_u^2 \sim N\left(\gamma_0, \left(\frac{\sigma_u^2}{E[\sigma_u^2]}\right)\Phi_\gamma\right). \tag{B.22}$$

Pastor and Stambaugh recommended using diffuse or completely non-informative priors. They set $\sigma_\delta^2 = \infty$, or a very large number for computational purposes, making the prior mean δ_0 irrelevant and implying a diffuse prior for δ^i.

The values for s_0, ν_0, γ_0, and Φ_γ were set using empirical Bayesian approach based on cross-sectional moments:

- γ_0 and Φ_γ were set to the cross-sectional mean and variance of the γ^i estimates from the set of regressions (B.14) run for each fund $i = 1, ..., N$:

$$\gamma_0 = \frac{1}{N}\sum_{i=1}^{N}\hat{\gamma}^i \tag{B.23}$$

and

$$\Phi_\gamma = \frac{1}{N}\sum_{i=1}^{N}(\hat{\gamma}^i - \gamma_0)(\hat{\gamma}^i - \gamma_0)'. \tag{B.24}$$

- Since the inverted gamma distribution for σ_u^2 implied

$$E[\sigma_u^2] = \frac{\nu_0 s_0^2}{\nu_0 - 2} \tag{B.25}$$

and

$$\nu_0 = 4 + \frac{2(E[\sigma_u^2])^2}{Var[\sigma_u^2]}, \tag{B.26}$$

the cross-sectional mean and variance of $\hat{\sigma}_u^2$ could be used to calculate ν_0 and s_0. Given

$$\widehat{E[\sigma_u^2]} = \overline{\sigma}_u^2 = \frac{1}{N}\sum_{i=1}^{N}\hat{\sigma}_{u^i}^2 \tag{B.27}$$

and

$$\widehat{Var[\sigma_u^2]} = \frac{1}{N}\sum_{i=1}^{N}\left(\hat{\sigma}_{u^i}^2 - \overline{\sigma}_u^2\right)^2 \tag{B.28}$$

Pastor and Stambaugh recommended setting ν_0 to the next largest integer after the value calculated using expression (B.26) and then using that value of ν_0 to calculate s_0 using expression (B.25).

Typically hedge funds are evaluated using an OLS approach with a fixed rolling window S approach (e.g., S = 24, 36, or 60). For each hedge fund i, an OLS regression (B.12) is applied using its returns $r^i = (r^i_{T-S+1}, ..., r^i_T)'$ for the period between $T - S + 1$ and T. The t-statistic of alpha $\alpha^i/\sigma(\alpha_i)$ is the standard approach to evaluating hedge funds. It is often measured using a Newey-West adjustment for heteroskedasticity and serial correlation. This adjustment has no impact on the estimates of α but it tends to reduce the value of the t-statistic of alpha when returns exhibit positive serial correlation and heteroskedasticity.

The implementation steps of the Pastor-Stambaugh approach are outlined below:

1. **Step 1: estimate parameters related to the hedge fund benchmarks using all available data for $t = 1, ..., T$.** Once the hedge fund benchmarks are regressed relative to the factors as in (B.13), the posterior distributions of the benchmark parameters as calculated as follows:

   - Set s^2 to the average of the diagonal elements of the covariance matrix of the residuals.
   - Set $\sigma^2_{\alpha_F} = 10^{15}$, or another sufficiently large number to represent a diffuse prior that can be handled computationally.
   - Set $Z = (\overline{1}_T F)$.
   - Set $D = 0$, an $(M+1)$-by-$(M+1)$ matrix of zeros, and then redefine the top left element $D_{1,1} = s^2/\sigma^2_{\alpha_F}$.
   - Set $W = D + Z'Z$.
   - Denote $G = (\alpha_F \beta^F)'$, where $\alpha_F = (\alpha^1_F, ..., \alpha^M_F)'$ and

$$\beta^F = \begin{pmatrix} \beta^1_{F,1} & \beta^1_{F,2} & \cdots & \beta^1_{F,K} \\ \beta^2_{F,1} & \beta^2_{F,2} & \cdots & \beta^2_{F,K} \\ \vdots & \vdots & \ddots & \vdots \\ \beta^M_{F,1} & \beta^M_{F,2} & \cdots & \beta^M_{F,K} \end{pmatrix} \tag{B.29}$$

are the intercept vector and slope matrix from expression (B.13).

- Set $\hat{G} = (Z'Z)^{-1} Z'F$ and $\hat{g} = vec(\hat{G})$.
- Set $\hat{\Sigma} = (F - Z\hat{G})'(F - Z\hat{G})/T$.
- Set $\nu = M+3$, the degrees of freedom in the inverted Wishart distribution, which serves as the prior for the covariance matrix Σ.
- Set $H = s^2 (\nu - M - 1) I_M$.
- Set $Q = Z' \left(I_T - Z (W)^{-1} Z'\right) Z$.
- The posterior estimate of the covariance matrix

$$\tilde{\Sigma} = \frac{1}{T + \nu - M - K - 1} \left(H + T\hat{\Sigma} + \hat{G}'Q\hat{G}\right). \tag{B.30}$$

- The posterior estimate $\tilde{g} = (I_M \otimes W^{-1}Z'Z)\hat{g}$.
- The posterior variance of g

$$\widetilde{Var}(g) = \tilde{\Sigma} \otimes W^{-1}. \tag{B.31}$$

- Since $G = (\alpha_F \beta^F)'$ and $g = vec(G)$, the posterior estimate

$$\tilde{\alpha}_F = \tilde{g}_{1,\ldots,M} = \begin{pmatrix} \tilde{g}_1 \\ \tilde{g}_2 \\ \vdots \\ \tilde{g}_M \end{pmatrix}, \tag{B.32}$$

the top M elements of $\tilde{g}$.
Its covariance matrix V_{α_F} is the M-by-M top left corner of the $\widetilde{Var}(g)$:

$$V_{\alpha_F} = \begin{pmatrix} \widetilde{Var}(g)_{1,1} & \widetilde{Var}(g)_{1,2} & \cdots & \widetilde{Var}(g)_{1,M} \\ \widetilde{Var}(g)_{2,1} & \widetilde{Var}(g)_{2,2} & \cdots & \widetilde{Var}(g)_{2,M} \\ \vdots & \vdots & \ddots & \vdots \\ \widetilde{Var}(g)_{M,1} & \widetilde{Var}(g)_{M,2} & \cdots & \widetilde{Var}(g)_{M,M} \end{pmatrix}. \tag{B.33}$$

2. **Step 2: For each hedge fund i its alpha and t-statistic of alpha are estimated using the rolling window S with t between $T - S + 1$ and T.** The estimation procedure starts

with deriving posterior estimates of the regression (B.14) and then using those estimates to estimate the alpha and the t-statistic of alpha using equation (B.16), $\alpha^i = \delta^i + \sum_{m=1}^{M} \gamma^i_{F,m}\alpha^m_F$, which can be re-written in the vector form as

$$\alpha^i = (\phi^i)'d, \tag{B.34}$$

where $\phi^i = (\delta^i\gamma^i)'$ and $d = (1\alpha_F'0...0)'$ with $\alpha_F = (\alpha^1_F...\alpha^M_F)'$. Since $\gamma^i = (\gamma^i_{F,1},...,\gamma^i_{F,M},\gamma^i_{f,1},...,\gamma^i_{f,K})'$, there are exactly K zeros at in the lower portion of the vector d.

The estimation is performed following the steps:

- As discussed previously, δ_0, the prior for δ is equal to zero, and the prior for γ_0 and Φ_γ are calculated as described in (B.23) and (B.24) based on the cross-sectional moments following the empirical Bayesian approach.
- Set $\phi_0 = (\delta_0\gamma_0')'$.
- The parameters of the prior inverted gamma distribution for σ^2_u are also calculated following the empirical Bayesian approach. The cross-sectional moments are calculated using (B.27) and (B.28) and then used to estimate ν_0 and s_0 as shown in expressions (B.25) and (B.26).
- Set

$$\Lambda_0 = \left(\frac{\nu_0 s_0^2}{\nu_0 - 2}\right)\begin{bmatrix}\sigma^2_\delta & 0\\ 0 & \Phi_\gamma\end{bmatrix}. \tag{B.35}$$

- Set Z^i to the last S rows of the matrix $Z = (\bar{1}_T F)$.
- The posterior mean of ϕ^i is equal to

$$\tilde{\phi}^i = (\Lambda_0 + Z_A'Z_A)^{-1}(\Lambda\phi_0 + (Z^i)'r^i). \tag{B.36}$$

and its variance

$$\widetilde{Var}(\phi^i) = V_{\phi^i} = \frac{h^i}{T+\nu_0-2}(\Lambda_0 + (Z^i)'Z^i)^{-1}, \tag{B.37}$$

where

$$h^i = \nu_0 s_0^2 + (r^i)'r^i + \phi_0'\Lambda_0\phi_0 - (\tilde{\phi}^i)'[\Lambda_0 + (Z^i)'Z^i]\tilde{\phi}^i.$$

- The posterior mean and covariance matrix of d are equal to

$$\tilde{d} = \begin{bmatrix} 1 \\ \tilde{\alpha}_F \\ 0 \end{bmatrix} \tag{B.38}$$

and

$$V_d = \begin{bmatrix} 0 & 0 & 0 \\ 0 & V_{\alpha_F} & 0 \\ 0 & 0 & 0 \end{bmatrix}. \tag{B.39}$$

- The posterior estimate of alpha

$$\tilde{\alpha}^i = (\tilde{\phi}^i)'\tilde{d} \tag{B.40}$$

and the posterior estimate of its variance is

$$\widetilde{Var}(\alpha^i) = tr\left(V_{\phi^i} V_d\right) + \tilde{d}' V_{\phi^i} \tilde{d} + (\tilde{\phi}^i)' V_d \tilde{\phi}^i. \tag{B.41}$$

The t-statistic of alpha is equal to the ratio of the alpha estimate and the square root of its variance.

Pastor and Stambaugh recommended using a small number of hedge fund benchmarks because if the number of benchmarks increases without a sufficient improvement in R^2 that would decrease rather than increase the precision of alphas.

B.5 IDENTIFYING HEDGE FUND SKILL WITH PEER COHORTS

This section includes the implementation details of the peer cohorts approach from the 2021 study *Identifying Hedge Fund Skill by Using Peer Cohorts* by David Forsberg, David Gallagher, and Geoffrey Warren.[18]

It is performed using the following steps:

1. **Forming cohorts with cluster analysis based on return correlations in excess of a threshold ρ_T.** Net returns are converted to gross returns using the method from the 2009 paper *Role of Managerial Incentives and Discretion in Hedge Fund Performance.*[19] Then the unweighted pair group method with arithmetic mean (UPGMA) technique is used to create hierarchical clusters. Each fund belongs to its own cluster, and then a hierarchical tree, or dendrogram, is created by sequentially linking together clusters with the closest similarity. For any two

funds i and j, the distance d_{ij} is equal to one minus their Pearson correlation ρ_{ij}:

$$d_{ij} = 1 - \rho_{ij}. \tag{B.42}$$

Note that the distance between any two funds is between 0 and 2. The distance between cluster A and B is defined as

$$d_{AB} = \frac{\sum\limits_{i \in A} \sum\limits_{j \in B} d_{ij}}{n(A)n(B)}, \tag{B.43}$$

where $n(A)$ and $n(B)$ represent the number of funds in clusters A and B, respectively.

Once again, the distance between any two clusters is between 0 and 2. Moreover, as the tree grows and the clusters sequentially are linked together, the number of clusters goes down and the average distance between funds within clusters goes up. The authors recommend the threshold distance value of 0.25, or the average pair-wise correlation of $\rho_T = 0.75$ of funds within clusters, to stop linking clusters. The resulting clusters are the cohorts used to evaluate funds.

The authors performed clustering quarterly using expanding rather than rolling windows of historical returns and required at least 24 months of returns for analysis. They also required having at least 2 funds in each cohort and repeated analysis using exponentially weighted correlations, different distance measures, and minimum cohort sizes to verify that their findings were robust.

2. **Calculating cohort alphas.**

For each fund i in cohort A that includes $n(A)$ funds, a customized benchmark f_t^i is calculated by averaging the returns of the remaining funds in the cohort:

$$f_t^i = \frac{1}{n(A) - 1} \sum_{j \neq i} r_t^j, \tag{B.44}$$

where r_t^j is return of a fund j from cohort A at time t.

Once the customized benchmark is calculated, α_C^i, the **cohort alpha** for fund i, is estimated using a standard ordinary least squares regression with a rolling 24-month window

$$r_t^i = \alpha_C^i + \beta_f^i f_t + \epsilon_t^i, \tag{B.45}$$

where β_f^i is the slope coefficient and ϵ_t^i is the residual.

B.6 COMBINING QUANTITATIVE AND QUALITATIVE FACTORS WITHIN A BAYESIAN FRAMEWORK

This section includes derivation of a threshold Sharpe value for selecting skilled hedge funds given the track record length and a proportion of skilled funds in the hedge fund pool. The role of qualitative due diligence is to increase the proportion of skilled funds in the pool.

An institutional investor can benefit from a simple rule that all hedge funds with a Sharpe ratio that exceeds a certain level are skilled funds with a high degree of confidence such as 95 percent, but it is unclear how the threshold level can be derived. It is intuitive to recognize two important factors:

- **The length of track record.** Because of the central limit theorem, the observed Sharpe ratios are closer to their true values when track records are longer.
- **The share of skilled and unskilled funds.** If most funds are skilled, then the threshold value can be very low. By contrast, if most funds are unskilled, the threshold value should be higher.

Denote T the track record length and p_S the share of skilled funds. Therefore, $1 - p_S$ is the shared of unskilled funds. Denote $f(S_T|S)$ and $f(S_T|U)$ as the probability density function of observing the Sharpe ratio of S_T for skilled and unskilled funds, respectively.

The Bayes theorem indicates that the likelihood of a fund with the Sharpe ratio S_T being a skilled fund can be calculated as

$$P(S|S_T) = \frac{p_S P(S_T|S)}{p_S P(S_T|S) + (1 - p_S)P(S_T|U)}, \tag{B.46}$$

where $P(S_T|S)$ and $P(S_T|U)$ represent the probability of the Sharpe ratio exceeding S_T for the skilled and unskilled funds, respectively.

For simplicity we assume normal distribution with $P(S_T|S) = 1 - F\left[(S_T - 1)\sqrt{T}\right]$ and $P(S_T|U) = 1 - F\left[S_T\sqrt{T}\right]$, where F is the standard cumulative distribution.

$$P(S|S_T) = \frac{p_S f(S_T|S)}{p_S f(S_T|S) + (1 - p_S)f(S_T|U)} = \\ \frac{p_S\left(1 - F\left[(S_T - 1)\sqrt{T}\right]\right)}{p_S\left(1 - F\left[(S_T - 1)\sqrt{T}\right]\right) + (1 - p_S)\left(1 - F\left[S_T\sqrt{T}\right]\right)}. \tag{B.47}$$

Since the institutional investor wants to set $P(S|S_T) = \alpha$ to a high probability such as $\alpha = 95\%$, the threshold level of S_T can be calculated numerically as a function of α and p_S.

Table B.1 **Threshold value of S_T as a function of T, the length of track record in years, and p_S, the proportion of skilled funds.**

T		2	3	5
	10.0%	2.8	2.0	1.4
	20.0%	2.3	1.7	1.2
	30.0%	2.2	1.5	1.0
p_S	40.0%	1.8	1.4	0.9
	50.0%	1.7	1.2	0.8
	60.0%	1.3	1.0	0.7
	70.0%	1.2	0.7	0.6

Table B.1 displays threshold values of S_T given the confidence level of 95 percent. It shows why the role of qualitative factors highlighted in section 3.2.6 is so significant. If a hedge fund has qualitative weakness that suggest that there is only 10 percent of similar funds that are skilled, it would require a 5-year Sharpe ratio of 1.4 to have sufficient confidence in the fund. By contrast, if a fund is qualitatively very strong indicating that 70 percent of such funds are skilled, it may require a much smaller Sharpe ratio of 0.7 over shorter period of 3 years to determine that the fund is skilled.

APPENDIX C

From Mean Variance to Risk Parity

C.1 SHRINKAGE ESTIMATORS

This section includes a thorough discussion of shrinkage estimators and the implementation details of shrinkage estimators for mean and covariance matrix.

In the 1956 study *Inadmissibility of the Usual Estimator for the Mean of a Multivariate Normal Distribution,* Charles Stein showed that a sample mean was a poor estimator for the mean of a multivariate normal distribution when a quadratic loss function was considered.[1]

While the formal derivation is complex, the finding has an intuitive explanation when considering two properties of estimators in one-dimensional case: bias, defined as the difference between the expectation of the estimator $E[\hat{\theta}]$ and the true value θ as $B(\hat{\theta}) = E[\hat{\theta}] - \theta$, and efficiency defined as the variance of the estimator $V(\hat{\theta}) = var(\theta)$. The sample mean $\overline{\theta}$ is an unbiased estimator that is not efficient. A simple constant value θ_0, called a target value, has zero variance (very efficient), but can potentially have a high bias. However, since the quadratic error function is related to $B^2(\hat{\theta}) + V(\hat{\theta})$, then $(1 - w)\hat{\theta} + w\theta_0$, a weighted average of the sample mean $\overline{\theta}$ and the constant value θ_0 can have lower overall error if the weight coefficient w is selected correctly.

Indeed, if $ERROR(\hat{\theta}) = B^2(\hat{\theta}) + V(\hat{\theta})$ and the variance of the sample mean estimator, which is unbiased, is equal to σ^2, then the error of the sample mean $ERROR(\overline{\theta})$ is equal to σ^2. Since the variance of the constant value estimator is zero, its error is purely driven by the bias $ERROR(\theta_0) = (\theta_0 - \theta)^2$.

DOI: 10.1201/9781003175209-C

The error of the shrinkage estimator is equal to

$$f(w) = ERROR((1-w)\hat{\theta} + w\theta_0) = (1-w)^2\sigma^2 + w^2(\theta_0 - \theta)^2. \quad \text{(C.1)}$$

The First Order Condition with respect to w is equal to $2(w-1)\sigma^2 + 2w(\theta_0 - \theta)^2 = 0$ producing the optimal weight value of

$$w_s = \frac{\sigma^2}{\sigma^2 + (\theta_0 - \theta)^2}. \quad \text{(C.2)}$$

This result makes intuitive sense. If the variance σ^2 is high, the sample mean is very inefficient and there is a lot of benefit from reducing it by using a higher value of w_s to shrink the estimator toward a constant value. If the target value is poorly chosen, its bias squared $(\theta_0 - \theta)^2$ is high and, therefore, there is less benefit from shrinking toward that target, which is consistent with a lower value of w_s.

It is convenient to note that $1 - w_s = \frac{(\theta_0-\theta)^2}{\sigma^2+(\theta_0-\theta)^2}$. The optimal value of w from expression (C.2) used within the error function (C.1) produces the value

$$f(w_s) = (1-w_s)^2\sigma^2 + w_s^2(\theta_0 - \theta)^2 =$$

$$\left(\frac{(\theta_0 - \theta)^2}{\sigma^2 + (\theta_0 - \theta)^2}\right)^2 \sigma^2 + \left(\frac{\sigma^2}{\sigma^2 + (\theta_0 - \theta)^2}\right)^2 (\theta_0 - \theta)^2 =$$

$$\frac{1}{(\sigma^2 + (\theta_0 - \theta)^2)^2}\left[\sigma^2(\theta_0 - \theta)^2(\sigma^2 + (\theta_0 - \theta)^2)\right] =$$

$$\frac{(\theta_0 - \theta)^2}{\sigma^2 + (\theta_0 - \theta)^2}\sigma^2 < \sigma^2. \quad \text{(C.3)}$$

Thus, a shrinkage estimator has lower error than a sample mean.

We illustrate this striking characteristic of shrinkage estimators using a hypothetical example shown in Table C.1.

The sample mean estimator is associated with a variance of 0.01. We also consider two potential targets. The first one is a mediocre target that has a bias of 20% = 0.2 and the bias squared of $0.2^2 = 0.04$. The second target is very attractive with a much smaller bias of 5% = 0.05 and the bias squared of $0.05^2 = 0.0025$.

Figure C.1 shows the bias squared, variance and the error of weighted averages of the sample mean estimator and the target estimator. When the weight of the target estimator is equal to zero, the weighted average is equal to the sample mean. Its bias squared is equal to zero and

Table C.1 **A hypothetical example with shrinkage.** This table illustrates two potential targets: a mediocre one with a large bias and an attractive one with a small bias. The table shows the volatility σ and variance σ^2 of the sample mean estimator, the bias and the bias squared of the target value θ_0, the optimal shrinkage weight w_s, the error of the sample estimator and the shrinkage estimator, and the error reduction due to shrinkage.

	Mediocre	Attractive
σ	10%	10%
σ^2	0.0100	0.0100
$\lvert\theta - \theta_0\rvert$	20%	5%
$(\theta - \theta_0)^2$	0.0400	0.0025
w_s	0.2	0.8
Error of sample estimator	0.0100	0.0100
Error of shrinkage estimator	0.0080	0.0020
Error Reduction	20%	80%

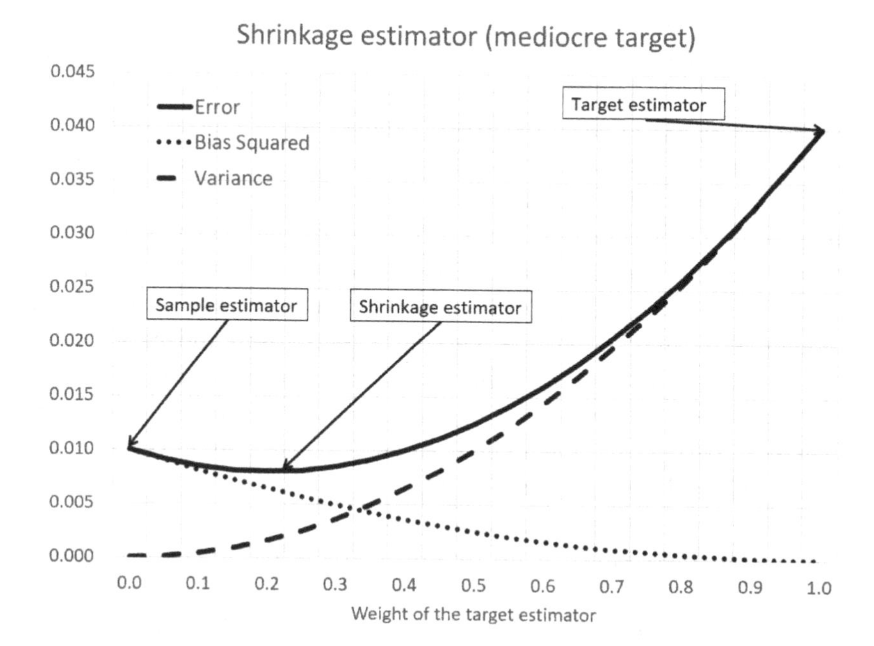

Figure C.1 Shrinkage estimator with a mediocre target.

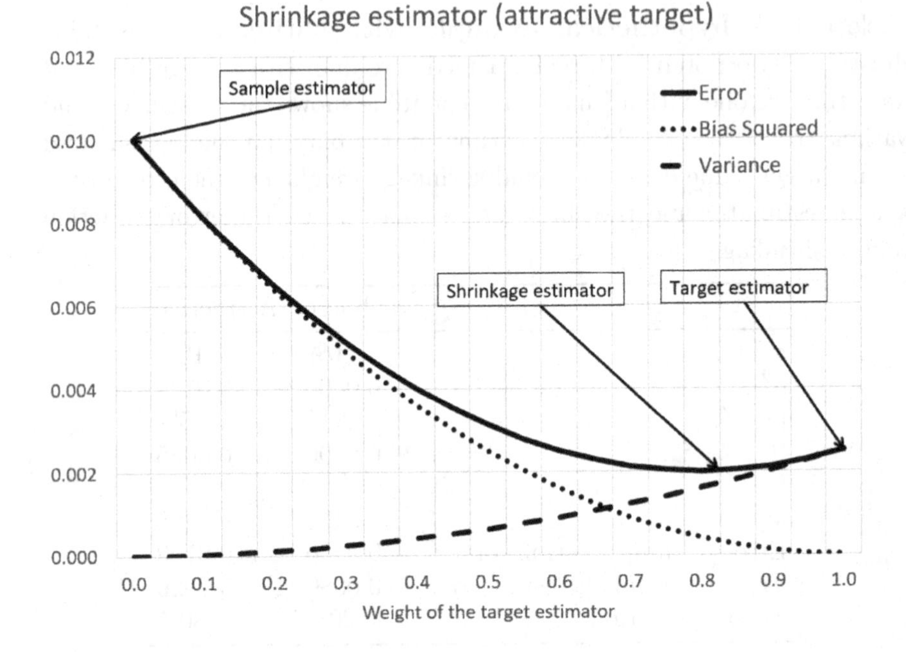

Figure C.2 Shrinkage estimator with an attractive target.

the variance is equal to 0.01. As the weight of the target estimator increases, the bias squared monotonically grows to 0.04 whereas the variance monotonically declines to zero. The optimal weight, calculated using expression (C.2), is equal to 0.2 and yields the error of 0.08, a modest 20% reduction in error relative to the sample mean estimator.

Figure C.2 shows the results for a more attractive target. It displays the bias squared, variance and the error of weighted averages of the sample mean estimator and the target estimator. When the weight of the target estimator is equal to zero, the weighted average is still equal to the sample mean with its bias squared equal to zero and its variance equal to 0.01. As the weight of the target estimator increases, the bias squared monotonically grows to 0.04 and the variance monotonically declines to zero. The optimal weight, calculated using expression (C.2), is equal to 0.8 and yields the error of 0.02, a striking 80% reduction in error relative to the sample mean estimator.

This result is striking because it illustrates a more general finding that a shrinkage estimator can be superior to the sample mean estimator for any target value θ_0. Of course, gains can be higher if the target value is chosen well and w_s should be estimated because the true value of θ is unknown.

The 1986 study *Bayes-Stein Estimation for Portfolio Analysis* by Philippe Jorion described the problem of estimating the mean vector of an N-dimensional normal random vector from an independent identically distributed time-series $\overline{y}_t = (y_{1t}, ..., y_{Nt})' \sim N(\overline{\mu}, \Sigma)$, where $t = 1, ..., T$ and Σ was considered normal as suggested by Robert Merton in the 1980 study *On Estimating the Expected Return on the Markets: An Exploratory Investigation.*[2,3]

Rather than using a sample mean $\overline{Y} = \sum_{t=1}^{T} y_t$, which is consistent with maximum likelihood estimation, Jorion suggested using the Bayes-Stein **shrinkage** estimator:

$$\hat{\mu}_{BS} = (1 - w)\overline{Y} + wY_0\overline{1}, \tag{C.4}$$

where weight given to the target Y_0

$$w = \frac{N + 2}{(N + 2) + (\overline{Y} - Y_0\overline{1})'T\Sigma^{-1}(\overline{Y} - Y_0\overline{1})} \tag{C.5}$$

and the target

$$Y_0 = \frac{\overline{1}'\Sigma^{-1}}{\overline{1}'\Sigma^{-1}\overline{1}}\overline{Y}. \tag{C.6}$$

Shrinkage estimation is often described as empirical Bayesian approach because it can be derived from the random means model using Bayesian estimation. The Bayes-Stein estimator has uniformly lower risk than the sample mean and results in superior out-of-sample performance relative to the sample mean when used within the mean-variance framework.

Olivier Ledoit and Michael Wolf in the 2003 study *Improved Estimation of the Covariance Matrix of Stock Returns with an Application to Portfolio Selection* and the 2004 study *Honey, I Shrunk the Sample Covariance Matrix* applied a similar shrinkage approach to improve the estimation quality for the covariance matrix.[4,5] They stated that "no one should use the sample covariance matrix for portfolio optimization" because shrinkage estimation systematically reduced estimation error when it mattered most. Conceptually, the idea of shrinkage was applied similarly to covariance matrices.

$$\Sigma_s = (1 - \delta)S + \delta F, \tag{C.7}$$

where S is the sample covariance matrix, F is the target and δ is the shrinkage weight, or shrinkage intensity as described by Ledoit and Wolf.

After S, the sample covariance matrix, was estimated, the estimation procedure of Ledoit and Wolf followed two steps: determining the target matrix F and estimating the shrinkage intensity δ.

First step: determining the target matrix F. The problem of shrinking covariance matrices is more nuanced than that of shrinking means because covariance matrices should be positive semi-definite, which requires imposing a specific structure on the target matrix F. The just cited 2003 study *Improved Estimation of the Covariance Matrix of Stock Returns with an Application to Portfolio Selection* used a single factor matrix suggested by William Sharpe in the 1963 study.[6] This approach works well for portfolios with a very strong common factor such as portfolios of equities. The aforementioned 2004 study *Honey, I Shrunk the Sample Covariance Matrix* suggested a constant correlation matrix approach where the target covariance matrix was constructed using sample variances of individual assets and all pair-wise correlations were equal to the average of the pair-wise sample correlations as follows.

A sample correlation ρ_{ij} between asset i and asset j was calculated from the sample covariance matrix S as $\rho_{ij} = \frac{s_{ij}}{\sqrt{s_{ii}s_{jj}}}$. Thus, the average sample correlation $\overline{\rho}$ could be calculated as

$$\overline{\rho} = \frac{2}{(N-1)N} \sum_{i=1}^{N-1} \sum_{j=i+1}^{N} \rho_{ij} = \frac{2}{(N-1)N} \sum_{i=1}^{N-1} \sum_{j=i+1}^{N} \frac{s_{ij}}{\sqrt{s_{ii}s_{jj}}}. \qquad \text{(C.8)}$$

The diagonal values of the target covariance matrix F were equal to the sample variances $f_{ii} = s_{ii}$ while all remaining values were calculated using sample variances and the average correlation $\overline{\rho}$ as $f_{ij} = \overline{\rho}\sqrt{s_{ii}s_{jj}}$.

Second step: estimating the shrinkage intensity. The process of estimating the shrinkage intensity δ_s was more complex. While the formula itself was simple

$$\delta_s = \max\left(0, \min\left(\frac{\kappa}{T}, 1\right)\right), \qquad \text{(C.9)}$$

the variable κ was non-trivial to estimate.

Indeed,

$$\kappa = \frac{\pi - \rho}{\gamma}, \qquad \text{(C.10)}$$

where parameter γ that measures the misspecification of the shrinkage target was calculated as

$$\gamma = \sum_{i=1}^{N}\sum_{j=1}^{N}(f_{ij} - s_{ij})^2, \tag{C.11}$$

parameter π, denoted the sum of asymptotic variances of the entries of the sample covariance matrix scaled by $\sqrt{T}$, calculated as follows

$$\pi = \sum_{i=1}^{N}\sum_{j=1}^{N}\pi_{ij} \tag{C.12}$$

with

$$\pi_{ij} = \frac{1}{T}\sum_{t=1}^{T}\left((y_{it} - \overline{Y}_i)(y_{jt} - \overline{Y}_j) - s_{ij}\right)^2. \tag{C.13}$$

Finally, ρ was the sum of asymptotic covariances of the entries of the shrinkage target with the entries of the sample covariance matrix scaled by $\sqrt{T}$:

$$\rho = \sum_{i=1}^{N}\pi_{ii} + \sum_{i=1}^{N}\sum_{j\neq i}\frac{\overline{\rho}}{2}\left(\sqrt{\frac{s_{jj}}{s_{ii}}}\nu_{ii,jj} + \sqrt{\frac{s_{ii}}{s_{jj}}}\nu_{jj,ii}\right) \tag{C.14}$$

where

$$\nu_{ii,jj} = \frac{1}{T}\sum_{t=1}^{T}\left((y_{it} - \overline{Y}_i)^2 - s_{ii}\right)\left((y_{it} - \overline{Y}_i)(y_{jt} - \overline{Y}_j) - s_{ij}\right). \tag{C.15}$$

As discussed in the 2016 paper *A New Diagnostic Approach to Evaluating the Stability of Optimal Portfolios*, a few popular target covariance matrices include assumptions of equal or zero correlations across the assets and equal or unequal variances across the assets.[7] The authors also suggested assessing the impact of shrinkage on the condition number of the covariance matrix as a simple diagnostic of the issue of instability of the mean-variance portfolio weights. If the condition number goes down, the shrinkage contributes to mitigating the instability.

In their 2017 study *Nonlinear Shrinkage of the Covariance Matrix for Portfolio Selection: Markowitz Meets Goldilocks*, Ledoit and Wolf introduced a nonlinear shrinkage estimator that was more flexible than their linear shrinkage estimators from their 2003 and 2004 studies.[8,9,10]

The authors showed that the optimal number of parameters within the estimator was equal to the number of assets and demonstrated that the nonlinear shrinkage estimator outperformed linear estimators for large stock portfolios with the number of portfolio constituents comparable to the sample size. The 2022 paper *The Power of (Non-)Linear Shrinking: A Review and Guide to Covariance Matrix Estimation* by Ledoit and Wolf provided a comprehensive overview of linear and nonlinear shrinkage approaches and demonstrated how nonlinear shrinkage could further improve performance when overlaid with stylized facts such as time-varying co-volatility or factor models.[11] The 2021 study *Shrinkage Estimation of Large Covariance Matrices: Keep It Simple, Statistician?* gave additional empirical support for the nonlinear shrinkage methodology using Monte Carlo simulations.[12]

C.2 BLACK-LITTERMAN OPTIMIZATION

This section describes details of Black-Litterman optimization from the 1992 study *Global Portfolio Optimization* by Fischer Black and Robert Litterman.[13]

First step: a prior portfolio. As discussed by Black and Litterman, the starting portfolio should be consistent with the "neutral" perspective (i.e., having no views or feelings whether some assets are overvalued or undervalued at current market prices), in which all investors have identical views regarding the set of expected returns. Thus, the starting portfolio should clear the market and, therefore, that starting portfolio should be the market portfolio as the equilibrium theory would suggest.

Second step: reverse optimization. Since the market portfolio is optimal for a mean-variance investor who optimizes

$$w_m = \arg\max_{w} \left[\pi_m' w - \frac{\gamma}{2} w' \Sigma w\right], \tag{C.16}$$

where γ is the risk-aversion coefficient, π_M is the vector of expected excess returns, and Σ is the covariance matrix of excess returns that is considered to be known, or can be easily estimated as shown in the 1980 study *On Estimating the Expected Return on the Markets: An Exploratory Investigation*.[14]

Since the solution of equation (C.16) results in $w_m = \frac{1}{\gamma}\Sigma^{-1}\pi_m$, the vector of expected excess returns implied by the market weights w_m can be recovered by performing reverse optimization

$$\pi_m = \gamma \Sigma w_m. \tag{C.17}$$

Although this step is easy quantitatively, it is quite remarkable because it enables investors to assess whether market expectations are appropriate. Indeed it is nearly impossible to determine whether a 2 percent allocation to a commodity index is reasonable whereas it is more tractable to draw conclusions whether the implied expected excess return of the commodity index of 25 percent per annum is reasonable. Similarly, it is very challenging to critique a 3 percent allocation to the Euro (EUR/USD) and a 4 percent allocation to the British Pound (GBP/USD), but it is possible to assess whether an expected return of 8 percent per annum for the British Pound is too high relative to the expected return of 1 percent for the Euro.

Third step: expert opinions. Investors often have forward-looking views about absolute or relative performance of assets. The Black-Litterman can incorporate investor views or expert opinions as long as they can be translated into statements regarding expected returns of assets with some degree of uncertainty as

$$P\pi \sim N(\nu, \Omega), \tag{C.18}$$

where P is the matrix of opinions, ν is the vector of opinions and Ω is the matrix of uncertainty regarding the opinions.

Let us illustrate how expert opinions can be expressed within Black-Litterman opinion using the following two examples. For convenience we assume the global portfolio includes N assets with the index commodity, the British Pound and the Euro listed as assets 1, 2, and 3, respectively.

1. **An absolute view.** The expected excess return of the commodity index is 5 percent with the standard deviation of that view of 10 percent. In other words, the 90 percent confidence interval for the expected excess return of commodities is between $5\% - 1.65 * 10\% \approx -11.5$ percent and $5\% + 1.65 * 10\% \approx 21.5$ percent. The idea of uncertainty is important because it mitigates the instability issue that has turned mean-variance optimization into a beautiful theory with ugly results. This view can be expressed within the Black-Litterman framework with $p_1 = [1, 0, ..., 0]$, $\nu_1 = 5$ percent and $\Omega_{11} = (10\%)^2 = 0.01$.

2. **A relative view.** The expected excess return of the British Pound and the Euro are the same with the standard deviation of the view of 2 percent. In this case, an investor has no opinion whether the British Pound and the Euro will go up or down, but the investor thinks that they should go up or down by about the same amount with the 90 percent confidence interval for the relative performance between $-1.65 * 2\% \approx -3.3$ percent and $+3.3$ percent. This view can be expressed within the Black-Litterman framework with $p_2 = [0, 1, -1, 0, ..., 0]$, $\nu_2 = 0$ percent and $\Omega_{22} = (2\%)^2 = 0.0004$.

Assuming that the opinions are independent, in which case the non-diagonal coefficients of Ω are equal to zero, we get the following representation of the two views within the Black-Litterman framework:

$$\begin{bmatrix} 1 & 0 & 0 & 0 & ...0 \\ 0 & 1 & -1 & 0 & ...0 \end{bmatrix} \pi \sim N\left(\begin{bmatrix} 5\% \\ 0\% \end{bmatrix}, \begin{bmatrix} 0.01 & 0 \\ 0 & 0.0004 \end{bmatrix} \right) \tag{C.19}$$

Forth step: a posterior portfolio. The final step assumes that

- The market equilibrium and the expert opinions provide two distinct sources of information regarding future expected excess returns.

- Both sources are associated with some uncertainty. As we have already discussed, the expert opinions are expressed as $P\pi \sim N(\nu, \Omega)$. The equilibrium excess returns are assumed to be also normally distributed with $N(\pi_m, \tau\Sigma)$, where π_m is the solution of reverse optimization shown in equation (C.17), Σ is the covariance matrix of the assets and τ is the uncertainty parameters. As discussed in the paper, the parameter τ should be very small because the uncertainty in the mean is much smaller than the uncertainty in the return itself.

- The posterior expected excess returns should be as consistent as possible with both sources of information. Applying Bayesian statistics produces the following expression for the vector of expected excess returns:

$$\pi_{BL} = \pi_m + \tau\Sigma P' \left(P\tau\Sigma P' + \Omega\right)^{-1} \left(\nu - P\pi_m\right). \tag{C.20}$$

The parameter τ is often set to $1/T$ and the posterior asset allocation can be calculated as

$$w_{BL} = \frac{1}{\gamma}\Sigma^{-1}\pi_{BL}. \tag{C.21}$$

If we consider a special case of the complete confidence in the expert opinions, the posterior vector of expected excess returns can be found using the constrained minimization

$$\pi_{BL} = \underset{P\pi=\nu}{\arg\min} \, (\pi - \pi_m) \left[\tau\Sigma\right]^{-1} (\pi - \pi_m)'. \tag{C.22}$$

This constrained optimization generates the conditional distribution for the expected returns π subject to the restrictions expressed in the expert opinions with the following mean:

$$\pi_{BL} = \pi_m + \tau\Sigma P' \left(P\tau\Sigma P'\right)^{-1} (\nu - P\pi_m). \tag{C.23}$$

As shown above, the Black-Litterman optimization beautifully applies Bayesian statistics to combine two seemingly contradictory ideas: the efficiency of the market portfolio and the benefit of expert opinions that may discover inefficiencies that are hidden to the market participants. Black-Litterman optimization is a popular portfolio technique used by practitioners.

C.3 HOW INEFFICIENT IS $1/N$?

This section describes a derivation of the estimation window required for mean-variance optimization from the 2009 study *Optimal versus Naive Diversification: How Inefficient is the 1/N Portfolio Strategy?* by Victor DeMiguel, Lorenzo Garlappi, and Raman Uppal.[15]

In the 2007 paper *Optimal Portfolio Choice with Parameter Uncertainty,* Raymond Kan and Guofu Zhou discussed the concept of an expected loss in utility U from using a particular estimator $\hat{w}$ of the true optimal portfolio weights w as[16]

$$L(w, \hat{w}) = U(w) - E[U(\hat{w})]. \tag{C.24}$$

Kan and Zhou showed that when both the mean and the covariance matrix were unknown, for a given window M used for estimation of the standard sample-based mean-variance estimator of portfolio weights $\hat{w}_{mv}$, the expected loss was equal to

$$L(w, \hat{w}_{mv}) = \frac{1}{\gamma}\left[(1-k)S_T^2 + h\right], \tag{C.25}$$

where γ is the coefficient of risk aversion, S_T is the Sharpe ratio of the true tangency portfolio,

$$k = \left(\frac{M}{M-N-2}\right)\left(2 - \frac{M(M-2)}{(M-N-1)(M-N-4)}\right) < 1, \tag{C.26}$$

and

$$h = \frac{NM(M-1))}{(M-N-1)(M-N-2)(M-N-4)} > 0. \qquad \text{(C.27)}$$

DeMiguel, Garlappi, and Uppal showed that the expected loss from relying on the naive 1/N approach with equal-weighted weights $\hat{w}_{ew}$ was equal to

$$L(w, \hat{w}_{ew}) = \frac{1}{\gamma}\left[S_T^2 - S_{ew}^2\right], \qquad \text{(C.28)}$$

where S_{ew} was the Sharpe ratio of the 1/N portfolio.

Therefore, the sample-based mean-variance portfolio should outperform the naive 1/N approach if $L(w, \hat{w}_{mv}) < L(w, \hat{w}_{ew})$, or

$$\frac{1}{\gamma}\left[(1-k)S_T^2 + h\right] < \frac{1}{\gamma}\left[S_T^2 - S_{ew}^2\right], \qquad \text{(C.29)}$$

which got simplified to the condition inequality

$$kS_T^2 - S_{ew} - h > 0. \qquad \text{(C.30)}$$

Thus, the critical window $M_c = \min(M : kS_T^2 - S_{ew} - h > 0)$ is a function of three variables: the number of assets, the true ex-anti Sharpe ratio of the mean-variance efficient portfolio and the Sharpe ratio of the 1/N portfolio. The critical window M_c has

- A positive relation with the number of assets. Since a bigger number of assets requires estimating a bigger number of parameters, the estimation error is higher, and, therefore, a longer window is required to sufficiently reduce the estimation error.

- A negative relation with the true ex-anti Sharpe ratio of the mean-variance efficient portfolio.

- A positive relation with the Sharpe ratio of the 1/N portfolio.

When calibrating the model to U.S. stock-market data, DeMiguel, Garlappi, and Uppal found that the critical window was 3,000 months for a portfolio with 25 assets and more than 6,000 for a portfolio with 60 assets. This finding questions the common practice of using 60-120 month windows in portfolio optimization.

C.4 NAIVE $1/N$, MINIMUM-VARIANCE, OR EQUAL RISK?

This section uses a simple example of two uncorrelated assets A and B to derive the conditions when the naive $1/N$, the minimum-variance, or the equal risk approach are optimal.

In this example, σ_A, volatility of the asset A, is equal to 10 percent and σ_B, volatility of the asset B, is equal to 20 percent. We also assume that μ_A, the expected excess return of asset A, is equal to 5 percent. For which value of μ_B would the 1/N, minimum-risk or equal-risk approach will produce the optimal highest Sharpe portfolio?

Since the portfolio weights w_s that maximize the Sharpe ratio of a portfolio of assets with the vector of expected excess returns μ and the covariance matrix Σ are equal to $\frac{\Sigma^{-1}\mu}{1'\Sigma^{-1}\mu}$ and the assets are uncorrelated, the optimal portfolio weights are equal to

$$w_A = \frac{\frac{\mu_A}{\sigma_A^2}}{\frac{\mu_A}{\sigma_A^2} + \frac{\mu_B}{\sigma_B^2}} \tag{C.31}$$

for asset A and

$$w_B = \frac{\frac{\mu_B}{\sigma_B^2}}{\frac{\mu_A}{\sigma_A^2} + \frac{\mu_B}{\sigma_B^2}} \tag{C.32}$$

for asset B.

Thus, the 1/N approach will produce the optimal Sharpe portfolio if $w_A = w_B$, which is equivalent to

$$\frac{\mu_A}{\sigma_A^2} = \frac{\mu_B}{\sigma_B^2}. \tag{C.33}$$

In our simple example that would imply

$$\mu_B = \mu_A \frac{\sigma_B^2}{\sigma_A^2} = 5\% \frac{(20\%)^2}{(10\%)^2} = 20\%. \tag{C.34}$$

Thus, if assets are uncorrelated, the 1/N approach is optimal if expected excess returns of assets are proportional to their variances.

Since the assets A and B are uncorrelated, the minimum-variance approach minimizes $w'\Sigma w$, which is equal to $w_A^2\sigma_A^2 + w_B^2\sigma_B^2$. The First Order Conditions produce the minimum-variance weights

$$w_A = \frac{\frac{1}{\sigma_A^2}}{\frac{1}{\sigma_A^2} + \frac{1}{\sigma_B^2}} \tag{C.35}$$

for asset A and

$$w_B = \frac{\frac{1}{\sigma_B^2}}{\frac{1}{\sigma_A^2} + \frac{1}{\sigma_B^2}} \tag{C.36}$$

for asset B.

Thus, the minimum-variance approach will produce the optimal Sharpe portfolio if $w_A = w_B$, which is equivalent to

$$\mu_A = \mu_B. \tag{C.37}$$

In our simple example that would imply $\mu_B = \mu_A = 5$ percent. Thus, if assets are uncorrelated, the minimum-variance approach is optimal if expected excess returns are the same across assets.

Finally, the equal volatility adjusted approach, described in the 2012 study *A Proof of Optimality of Volatility Weighting over Time,* also known as the inverse volatility approach because asset weights are inversely related to volatility:

$$w_A = \frac{\frac{1}{\sigma_A}}{\frac{1}{\sigma_A} + \frac{1}{\sigma_B}} \tag{C.38}$$

for asset A and

$$w_B = \frac{\frac{1}{\sigma_B}}{\frac{1}{\sigma_A} + \frac{1}{\sigma_B}} \tag{C.39}$$

for asset B.[17]

Thus, the equal risk approach will produce the optimal Sharpe portfolio if $w_A = w_B$, which is equivalent to

$$\frac{\mu_A}{\sigma_A} = \frac{\mu_B}{\sigma_B}. \tag{C.40}$$

In our simple example that would imply

$$\mu_B = \mu_A \frac{\sigma_B}{\sigma_A} = 5\% \frac{20\%}{10\%} = 10\%. \tag{C.41}$$

Thus, if assets are uncorrelated, the equal risk approach is optimal if the Sharpe ratios are the same across assets.

C.5 RISK PARITY

This section uses a simple example of two uncorrelated assets A and B to derive the conditions when the naive $1/N$, the minimum-variance, or the equal risk approach are optimal.

C.5.1 Standard Risk Parity: Equal Risk Contribution

As discussed for a more general case in detail in Section C.5.2, as a homogeneous function of order one, portfolio volatility $\sigma_p = \sqrt{w'\Sigma w}$ can be expressed as:

$$\sigma_p(w) = \sum_{i=1}^{N} \left[\frac{\partial \sigma_p}{\partial w_i} w_i \right] = \sum_{i=1}^{N} TRC_i^{\sigma}, \tag{C.42}$$

where $TRC_i^{\sigma} = \frac{\partial \sigma_p}{\partial w_i} w_i$ is the total risk contribution of asset i to portfolio volatility.

$MRC_i^{\sigma} = \frac{\partial \sigma_p}{\partial w_i}$, the marginal risk contribution of asset i to portfolio volatility and the total risk contribution can be expressed as $TRC_i^{\sigma} = MRC_i^{\sigma} w_i$.

The risk parity portfolio is the portfolio that has the same total risk contribution coming from each asset, or $TRC_i^{\sigma} = TRC_j^{\sigma}$ for each i and j. Imposing this condition on expression (C.42), results in

$$TRC_i^{\sigma} = \frac{1}{N} \sigma_p(w). \tag{C.43}$$

The 2006 study *On the Financial Interpretation of Risk Contribution: Risk Budgets Do Add Up* described PRC_i^{σ}, percentage contribution to risk from asset i, as[18]

$$PRC_i^{\sigma} = TRC_i^{\sigma} / \sigma = w_i \frac{\partial \sigma_p}{\partial w_i} / \sigma_p. \tag{C.44}$$

It follows from expression (C.42) that $\sum_{i=1}^{N} PRC_i^{\sigma} = 1$ and from expression (C.43) that the equal risk contribution portfolios have $PRC_i^{\sigma} = \frac{1}{N}$ for each asset i.

The percentage contribution to risk from asset i can be calculated as

$$PRC_i^{\sigma} = \frac{w_i \sigma_i \sum_{j=1}^{N} w_j \sigma_j \rho_{ij}}{\sigma_p^2}. \tag{C.45}$$

There is no general analytic solution for calculating portfolio weights of risk parity portfolios. However, if all correlations are equal to each other, $\rho_{ij} = \rho$, the equal risk contribution is equivalent to EVA, the equal volatility-adjusted allocation with weight of each asset i calculated as

$$w_i = \frac{1/\sigma_i}{\sum_{j=1}^{N} 1/\sigma_j} \tag{C.46}$$

and

$$w_i\sigma_i = \frac{1}{\sum_{j=1}^{N} 1/\sigma_j}. \tag{C.47}$$

Indeed, the portfolio variance of an equally volatility-adjusted portfolio is equal to

$$\sigma_p^2 = w'\Sigma w = \frac{1}{\left(\sum_{j=1}^{N} 1/\sigma_j\right)^2}(N + N[N-1]\rho) =$$

$$= \frac{N(1+[N-1]\rho)}{\left(\sum_{j=1}^{N} 1/\sigma_j\right)^2} \tag{C.48}$$

and the percentage risk contribution from asset i

$$PRC_i^{\sigma} = \frac{w_i\sigma_i \sum_{j=1}^{N} w_j\sigma_j\rho}{\sigma_p^2} = \frac{1}{\left(\sum_{j=1}^{N} 1/\sigma_j\right)^2} \frac{1+[N-1]\rho}{\sigma_p^2} =$$

$$= \frac{1}{\left(\sum_{j=1}^{N} 1/\sigma_j\right)^2} \frac{1+[N-1]\rho}{N(1+[N-1]\rho)} \left(\sum_{j=1}^{N} 1/\sigma_j\right)^2 = \frac{1}{N}. \tag{C.49}$$

Thus, if all pair-wise correlations are equal to each other, the equal risk contribution approach produces equal volatility-adjusted allocations.

C.5.2 A General Equal Risk Contribution Approach

While volatility is one of the most common measures of risk, the equal risk contribution idea works for any risk measure f, which is a homogeneous function of order one. f is a homogeneous function of order one if for any constant c and any vector of w the function satisfies the equation $f(cw) = cf(w)$.

The Euler's theorem for homogenous functions states that the function f can be expressed as

$$f(w) = \sum_{i=1}^{N} \left[\frac{\partial f}{\partial w_i} w_i\right]. \tag{C.50}$$

Similar to the case of classic parity, we can define $TRC_i^f = \frac{\partial f}{\partial w_i} w_i$ as the total risk contribution of asset i to risk measure f and re-write expression (C.50) as

$$f(w) = \sum_{i=1}^{N} \left[\frac{\partial f}{\partial w_i} w_i \right] = \sum_{i=1}^{N} TRC_i^f. \tag{C.51}$$

Then in a more general case, the equal risk contribution approach is defined by

$$TRC_i^f = \frac{\partial f}{\partial w_i} w_i = \frac{1}{N} f(w) \tag{C.52}$$

for any asset i.

Thus, the risk parity idea can be applied to popular risk measures of expected shortfall, also known as Conditional Value-at-Risk (CVaR) or Expected Tail Loss, maximum drawdowns or maximum loss. The expected shortfall is probably the most interesting example because it is highly regarded by regulators and practitioners as one of the best measures of risk. As discussed in the 1999 study *Coherent Measures of Risk,* this measure meets all the requirements of *coherent risk measures* including an obvious one of not penalizing diversification whereas Value-at-Risk (VaR) fails to do that.[19,a] In fact, the authors showed that VaR could behave poorly when even independent risks were aggregated and sometimes penalized diversification because it did not account for the economic consequences of the events, the probabilities of which it controlled.

The expected shortfall or CVaR is defined as the expected "average" loss beyond the VaR level. Let the portfolio returns r^p follow distribution F. The Value-at-Risk value given threshold value α is defined as $VaR_\alpha(r^p) = F^{-1}(\alpha)$. The expected shortfall is defined as

$$CVaR(r^p) = E[r^p | r^p < VaR_\alpha(r^p)]. \tag{C.53}$$

CVaR is a homogeneous function of order one. Indeed,

$$CVaR(cr^p) = E[cr^p | r^p < VaR_\alpha(r^p)] = \\ cE[r^p | r_p < VaR_\alpha(r^p)] = cCVaR(r^p). \tag{C.54}$$

[a]The 1999 study *Coherent Measures of Risk* defined coherent risk measures as risk measures that satisfy the four axioms of translation invariance, subadditivity, positive homogeneity, and monotonicity. The axiom of subadditivity directly relates to diversification because it states that the risk of a portfolio cannot exceed the sum of risks of its components.

At any point in time t, the portfolio return r_t^p can be expressed as a product of portfolio weights w and asset returns r_t. Thus, we can express $r_p = w'r$.

Therefore, according to the Euler's theorem for homogeneous functions

$$CVaR(r^p) = \sum_{i=1}^{N} \left[\frac{\partial CVaR(r^p)}{\partial w_i} w_i \right] = \sum_{i=1}^{N} TRC_i^{CVaR}. \tag{C.55}$$

The CVaR-based equal risk contribution approach is defined by

$$TRC_i^{CVaR} = \frac{\partial CVaR}{\partial w_i} w_i = \frac{1}{N} CVaR(r^p) \tag{C.56}$$

for any asset i and the portfolio weights can be found numerically.

The CVaR-based risk parity may be more applicable for hedge fund portfolios than the standard risk parity approach because it accounts for tail risk often present in hedge fund returns.

C.5.3 Risk Parity with Drawdowns

Investment decisions often require an evaluation of several measures of risk-adjusted performance and risk. The maximum historical drawdown is very popular because it represents the worst-case scenario for an investor who is unfortunate to lock such loss by investing at the top and redeeming at the bottom. Best practices in due diligence of alternative investments require drawdown analysis as part of standard quantitative due diligence.[20]

Therefore, it looks attractive to minimize that risk and build portfolios with low drawdowns. However, minimizing maximum historical drawdowns has three significant issues:

1. First, the issue of estimation error or overfitting for noisy realized returns without accounting for potential alternative outcomes tends to lead to poor out-of-sample performance as shown for mean-variance optimization in the 1986 paper *Bayes-Stein Estimation for Portfolio Analysis*, the 1989 study *The Markowitz Optimization Enigma: Is Optimized Optimal?*, and the 2009 paper *Optimal versus Naive Diversification: How Inefficient is the 1/N Portfolio Strategy?*[21,22,23]

2. Second, analytic solutions for maximum drawdowns are not feasible. The 2000 study *On Probability Characteristics of Downfalls in a Standard Brownian Motion* by Rafael Douady, Albert Shiryaev, and Marc Yor showed that analytic solutions for expected maximum drawdowns were limited to a Brownian motion with a drift.[24] Although their complex derivations are impressive, the Brownian motion with a drift is a poor approximation of financial time-series.

3. Finally, focusing on maximum historical drawdown, the worst-case scenario from within the drawdown distribution, is not robust and is likely to produce poor out-of-sample performance. The 2005 study *Drawdown Measure in Portfolio Optimization* by Alexei Chekhlov, Stanislav Uryasev, and Michael Zabarankin and the 2014 paper *On a Convex Measure of Drawdown Risk* by Lisa Goldberg and Ola Mahmoud proposed considering drawdown-based measures that were based on the left tail of the drawdown distribution rather than its single worst point.[25,26]

Chekhlov, Uryasev, and Zabarankin introduced a comprehensive framework for overcoming the issues of relying on historical maximum drawdowns. They proposed a family of risk measures called Conditional Drawdown (CDD), the tail mean of drawdown distributions, investigated its mathematical characteristics and discussed applications of the new measure to asset allocation decisions. The authors also introduced a block bootstrap procedure for the calculation of CDD and showed that the procedure was robust after approximately 100 simulations. A numerical approach is required as shown by Douady, Shiryaev, Marc Yor, and a block bootstrap approach preserves the serial and cross-correlational characteristics of the original dataset, which is important when constructing portfolios of hedge funds.

Goldberg and Mahmoud introduced the Conditional Expected Drawdown (CED), the expected tail of maximum drawdown distributions, defined as

$$CED^{\alpha}(\xi) = E\left[MDD(\xi) | MDD(\xi) < DT^{\alpha}(\xi)\right], \tag{C.57}$$

where $MDD(\xi)$ represents the maximum drawdown distribution of a random variable ξ, such as returns of a hedge fund portfolio, and $DT^{\alpha}(\xi)$ is the quantile of the maximum drawdown distribution that corresponds to probability α. This measure is very similar to CVaR defined in expression (C.53), but it applies to distributions of maximum drawdowns rather than raw returns.

Goldberg and Mahmoud showed that the CED was a coherent measure. Since it is a homogeneous function of order one, the Euler equation can once again serve to equalize total contributions of each asset to the portfolio CED within an CED-based risk parity approach:

$$CED^{\alpha}(r^{p}) = \sum_{i=1}^{N} \left[\frac{\partial CED^{\alpha}(r^{p})}{\partial w_i} w_i \right] = \sum_{i=1}^{N} TRC_i^{CED}. \tag{C.58}$$

The CED-based equal risk contribution approach is defined by

$$TRC_i^{CED} = \frac{\partial CED^{\alpha}}{\partial w_i} w_i = \frac{1}{N} CED^{\alpha}(r^{p}) \tag{C.59}$$

for any asset i and the portfolio weights can be found numerically. Goldberg and Mahmoud simplified the formulas for the total risk contribution TRC_i^{CED} to make it easier to construct a CED-based risk-parity portfolio within a simulation framework.

The CED measure effectively addresses two out of three issues highlighted at the beginning of this section: it considers the left tail of the distribution rather than the worst-case scenario and it uses numerical approaches instead of analytic solutions. However, it is unclear whether the CED approach is sensitive to sampling error and whether CED-based risk-parity produces attractive out-of-sample performance.

The 2017 study *Portfolio Management with Drawdown-Based Measures* showed that the CED-approach was sensitive to sampling error and its out-of-sample performance though superior to that based on maximum drawdown minimization was still poor.[27] One of the reasons why the CED was sensitive to sampling error could be because of its heavy reliance on the historical performance of assets via the slope of their cumulative return functions and their contributions to the portfolio performance during bad periods. The authors introduced a new drawdown risk metric, the Modified Conditional Expected Drawdown (MCED), which replaced asset returns with their de-meaned returns first and then applied the CED calculations using block bootstrap simulations. De-meaning mitigated the impact of the historical slope of cumulative returns functions and block bootstrap simulations attempted to reduce the impact of sampling error by evaluating what could happen given feasible alternative market environments.

The authors used simulations to demonstrate that their MCED approach was less sensitive to sampling error than the historical maximum drawdown and the CED approach of Goldberg and Mahmoud. They also showed that while within the large-scale simulation framework of Molyboga and L'Ahelec the MCED-based risk parity produced CTA portfolios with superior Sharpe or Calmar ratios than all drawdown-based techniques considered, it failed to consistently outperform the simple equal-volatility weighted approach.

C.6 ADAPTIVE OPTIMAL RISK BUDGETING

This section describes technical details of the adaptive optimal risk budgeting (AORB) approach from the 2020 study *Adaptive Optimal Risk Budgeting.*[28]

The approach is based on the finding from the 2001 paper *Implementing Optimal Risk Budgeting* and the 2006 study *The Sense and Nonsense of Risk Budgeting* that showed that RC_o, the optimal mean-variance risk budget vector, could be expressed as

$$RC_o = \frac{\Omega^{-1}S}{\sqrt{S'\Omega^{-1}S}} RC_T, \qquad \text{(C.60)}$$

where RC_T was the target risk budget, S was the vector of Sharpe ratios, and Ω was the correlation matrix.[29,30]

If the Sharpe ratios and the correlations are the same for all portfolio constituents, the equation (C.60) produces the classic risk-parity portfolio. If the Sharpe ratios and the correlations are estimated using historical data, the equation (C.60) produces the mean-variance solution.

The AORB approach assumes that the Sharpe ratios vary across portfolio constituents and across time:

$$S_t = \lambda S_{t-1} + (1-\lambda) S_t^h, \qquad \text{(C.61)}$$

where λ is a constant that is below but close to 1, S_t^h is the realized Sharpe ratio calculated using historical data and S_t is the forecasted Sharpe value, which is then used to calculate optimal risk budgets following equation (C.60). The AORB initially assumes that the Sharpe ratios are the same for all portfolio constituents but then learns from the historical data and adjusts the Sharpe ratios of the assets independently of each other following equation (C.61).

C.7 DIVERSIFICATION ACROSS TIME WITH VOLATILITY TARGETING

This section summarizes the theoretical results from the 2019 paper *Portfolio Management of Commodity Trading Advisors with Volatility Targeting* regarding conditions that should be satisfied for volatility targeting to improve the out-of-sample Sharpe ratio of a hedge fund portfolio.[31]

The study considered several versions of a theoretical model. In all cases the volatility was assumed to be known, since it is easy to estimate and highly persistent.[32,33] The author showed that if ex-anti Sharpe ratio was constant and correlations were constant, volatility targeting improved the expected Sharpe ratio. This result is intuitive because of diversification. Since the opportunity set, measured in terms of expected Sharpe ratio, is constant across time, the optimal risk allocation strategy is to maximize diversification across time by taking a constant risk exposure, which is accomplished by volatility targeting.

The author also allowed for the expected Sharpe ratio to vary with the level of volatility while keeping serial correlation equal to zero for simplicity. Specifically, they assumed a linear relationship between volatility and expected Sharpe ratio as follows

$$S_t = \hat{S} + \alpha(\widehat{\sigma}_t - \overline{\sigma}), \tag{C.62}$$

where $\overline{\sigma} = \frac{1}{T}\sum_{t=1}^{T}\widehat{\sigma}_t$ was the average level of volatility, $\hat{S} > 0$ was the average Sharpe ratio that was expected at the average level of volatility $\overline{\sigma}$, and α, the slope coefficient, could be either positive or negative.

The portfolio return was assumed to be equal to

$$r_t = \widehat{\sigma}_t(S_t + \epsilon_t), \tag{C.63}$$

where $\epsilon_t \sim N(0,1)$ was a standard normal variable and the returns were serially uncorrelated (i.e., $cor(\epsilon_t, \epsilon_\tau) = E(\epsilon_t\epsilon_\tau) = 0$ if $t \neq \tau$).

In this case, the optimal strategy was to vary risk and optimal scaling depended on the functional form of the link between volatility and expected Sharpe ratio (i.e., the values of α, $\hat{S}$, and the distribution of $\widehat{\sigma}_t$). However, since the study investigated the impact of volatility targeting rather than a potential benefit of performance timing, it compared the performance of the volatility managed strategy that targeted constant volatility to that of the base strategy that kept leverage constant.

The author found that the volatility managed strategy was generally expected to outperform except under a very strict set of conditions of a very strong positive relationship between volatility and expected Sharpe ratios that satisfied

$$\frac{S^+ - S^-}{\hat{S}} \geq \frac{2\sigma_\sigma}{\sqrt{\overline{\sigma}^2 + \sigma_\sigma^2} + \overline{\sigma}}, \tag{C.64}$$

where $\sigma_\sigma = \sqrt{\frac{1}{T}\sum_{t=1}^{T} \widehat{\sigma_t}^2 - \overline{\sigma}^2}$ was the standard deviation of volatility. $\hat{S}$, S^+, and S^- were the values of the Sharpe ratios at the average level of volatility $\overline{\sigma}$ and the levels of volatility at plus one σ_σ and minus one σ_σ, respectively.

As discussed in the paper, this inequality is very restrictive. If the average level of volatility, $\overline{\sigma}$, is equal to 10 percent and its volatility, σ_σ, is equal to 2 percent, then the ratio on the right hand side of the inequality is equal to 0.2. Thus, if the average Sharpe ratio, $\hat{S}$, is equal to 0.5, the difference in the Sharpe ratios associated with higher volatility and lower volatility periods has to be greater than 0.1. This positive difference is very significant in economic terms and inconsistent with previous empirical findings for assets. For example, the 2001 paper *The Specification of Conditional Expectations* found no evidence of a significant positive association between the stock market variance and expected return.[34] Furthermore, the 2007 study *Risk, Return and Dividends* suggested that the relationship could even be negative under plausible assumptions regarding the dynamics of expected returns and variance.[35]

While the impact of volatility targeting can vary across hedge fund strategies, the inequality (C.64) can serve as a rough diagnostic test to evaluate the potential impact. Moreover, the large-scale simulation framework of Molyboga and L'Ahelec can be used to evaluate the strategy given real life constraints.

APPENDIX D

Advanced Portfolio Construction

D.1 DENOISING CORRELATION MATRICES

This section includes technical details of the five approaches covered in the 2016 paper *Cleaning Correlation Matrices* by Joel Bun, Jean-Philippe Bouchaud, and Marc Potters.[1] The authors discussed four standard cleaning approaches (basic linear shrinkage, advanced linear shrinkage, eigenvalue clipping, and eigenvalue substitution) and proposed a new approach: rotationally invariant, optimal shrinkage.

The authors started with a portfolio with N constituents described by a vector of returns $r_t = (r_{1t}, ..., r_{Nt})$ for each $t = 1, ..., T$ (e.g., when monthly returns are used, T can be equal to 60). Correlation matrices are typically estimated after de-meaning returns of each portfolio constituent $\overline{r}_{it} = r_{it} - \mu_i$ and then standardizing returns using volatility estimates as $\tilde{r}_{it} = \overline{r}_{it}/\hat{\sigma}_{it}$.[a] As discussed in the 1980 paper *On Estimating the Expected Return on the Markets: An Exploratory Investigation,* volatility estimation is a tractable problem.[2]

The final adjustment was made by normalizing the standardized returns of each portfolio constituent by the sample estimator of their volatility $X_{it} = \tilde{r}_{it}/\sigma^i$.

[a]De-meaning can also be performed using auto-regressive models when the time-series of returns are autocorrelated.

DOI: 10.1201/9781003175209-D

The sample estimator of the true correlation matrix C was $E = \frac{1}{T}XX'$, which could be expressed as

$$E = \sum_{k=1}^{N} \lambda_k u_k u_k', \tag{D.1}$$

where the eigenvalues $\lambda_1 \geq \lambda_2 \geq ... \geq \lambda_N \geq 0$ with the corresponding eigenvectors $u_1, ..., u_N$. When $q = N/T \nrightarrow 0$, the large eigenvalue of E are too large and the small eigenvalues of E are too small when compared to the eigenvalues of C.

The authors described five cleaning approaches:

1. **Basic linear shrinkage**, a linear combination of the sample estimate and the identity matrix:

$$E^b = \alpha E + (1 - \alpha) I_N.$$

2. **Advanced linear shrinkage**, a linear combination of the sample estimate and the equal correlation matrix:

$$E^a = \alpha E + (1 - \alpha)(1 - \rho) I_N + \rho \overline{1}\,\overline{1}',$$

where $\overline{1}$ is an N-by-1 vector of ones and ρ is the average correlation suggested in the 2003 study *Improved Estimation of the Covariance Matrix of Stock Returns with an Application to Portfolio Selection.*[3]

3. **Eigenvalue clipping**, an approach that retains the top eigenvalues and shrinks the others to a constant γ that preserves the trace of E:

$$E^c = \sum_{k=1}^{N} \lambda_k^c u_k u_k',$$

where $\lambda_k^c = \lambda_k$, if $\lambda_k \leq N\alpha$, and $\lambda_k^c = \gamma$, otherwise.

4. **Eigenvalue substitution**, an approach that replaces the sample eigenvalues λ_k with estimates λ_k^{mp}, calculated using the Marcenko and Pastur equation:

$$E^c = \sum_{k=1}^{N} \lambda_k^{mp} u_k u_k'.$$

5. **Rotationally invariant, optimal shrinkage**, an approach that replaces the sample eigenvalues λ_k with optimal rotational invariant estimates λ_k^{RIE}:

$$E^c = \sum_{k=1}^{N} \lambda_k^{RIE} u_k u_k',$$

where

$$\lambda_k^{RIE} = \frac{\lambda_k}{|1 - q + q z_k s(z_k)|^2},$$

where $s(z) = \frac{1}{N} tr\left(zI_n - E\right)^{-1}$, tr is the matrix trace function, and $z_k = \lambda_k - i\nu$. ν is a small parameter that should also satisfy $N\nu >> 1$. One popular option is $\nu = \frac{1}{\sqrt{N}}$.

The authors showed that the advanced linear shrinkage and the eigenvalue substitution approaches were less efficient than the simpler basic linear shrinkage and eigenvalue clipping, respectively. When testing the remaining three approaches on empirical data, they found that the rotationally invariant, optimal shrinkage approach outperformed all other approaches out-of-sample. Thus, the authors recommended using the rotationally invariant estimators for large correlation matrices.

D.2 MEAN-VARIANCE EFFICIENCY FOR LARGE PORTFOLIOS

This section covers the implementation details of the maximum-Sharpe-ratio estimated and sparse regression ("MAXSER") approach from the 2018 study *Approaching Mean-Variance Efficiency for Large Portfolios.*[4]

Given N assets $r = (r^1, ..., r^N)$ with the mean vector of excess return μ and covariance matrix Σ, the standard mean-variance problem deals with maximizing expected portfolio return subject to the target portfolio volatility constraint σ_T:

$$w_{MV} = \underset{w:w'\Sigma w \leq \sigma_T^2}{\arg\max}\; w'\mu. \tag{D.2}$$

The authors showed that the mean-variance problem was equivalent to the unconstrained regression:

$$w_{MV} = \arg\min_{w} E[r_c - w'r]^2, \tag{D.3}$$

where $r_c = \frac{1+\theta}{\sqrt{\theta}}\sigma_T$ and $\theta = \mu'\Sigma^{-1}\mu$, the square of the Sharpe ratio of the optimal portfolio from (D.2).

The regression was replaced with the sample counterpart:

$$\hat{w}_{MV} = \arg\min_{w} \frac{1}{T}\sum_{t=1}^{T}\left(r_c - w' r_t\right)^2, \tag{D.4}$$

where $r_t = (r_t^1, ..., r_t^N)$.

As discussed in section 3.2.5, when N is large, standard statistical approaches produce unstable results and sparse regression techniques such as LASSO or Elastic Nets are required. The authors selected the LASSO approach:

$$\hat{w}_{MV} = \arg\min_{w:\sum_{i=1}^{N}|w_i|\leq\lambda} \frac{1}{T}\sum_{t=1}^{T}\left(r_c - w' r_t\right)^2. \tag{D.5}$$

The implementation steps of the MAXSER approach are outlined below:

1. **Step 1: estimate r_c.** The authors used the adjusted estimator of the maximum squared Sharpe ratio recommended by Kan and Zhou in their 2007 paper *Optimal Portfolio Choice with Parameter Uncertainty*:

$$\hat{\theta} = \frac{(T-N-2)\hat{\theta}_s - N}{T} + \frac{2(\hat{\theta}_s)^{N/2}(1+\hat{\theta}_s)^{-(T-2)/2}}{TB_{\hat{\theta}_s/(1+\hat{\theta}_s)}(N/2,(T-N)/2)} \tag{D.6}$$

where $\hat{\theta}_s = \hat{\mu}'\hat{\Sigma}^{-1}\hat{\mu}$. $\hat{\mu}$ and $\hat{\Sigma}$ were the sample mean and covariance matrix, respectively, and[5,b]

$$B_x(a,b) = \int_0^x y^{a-1}(1-y)^{b-1}dy.$$

Then, r_c was estimated as:

$$\hat{r}_c = \sigma_T \frac{1+\hat{\theta}}{\sqrt{\hat{\theta}}}, \tag{D.7}$$

where $\hat{\theta}$ was defined in equation (D.6).

[b]The 2007 paper *Optimal Portfolio Choice with Parameter Uncertainty* presented the formula of the unbiased estimator of the square of maximum Sharpe ratio $\hat{\theta} = \frac{(T-N-2)\hat{\theta}_s - N}{T}$, but found that it often took negative values and proposed an adjustment.

2. **Step 2: estimate optimal weights for λ that varies between 0 and λ_{OLS}, the L_1 norm of the ordinary least squares solution of (D.4).** The authors noted that the values of λ that were larger than λ_{OLS} produced the portfolio weights that matched the OLS solution. The least angle regression (LARS) approach from the 2004 paper *Least Angle Regression* efficiently solved for the entire solution path that included an estimated optimal portfolio for each value of λ.[6]

3. **Step 3: estimate the tuning parameter λ.** The authors recommended using a standard 10-fold cross-validation approach to select the value $\hat{\lambda}$ that corresponded to an optimal portfolio with volatility that was closest to σ_T. They reported that their simulations and empirical analysis confirmed that the cross-validation procedure produced accurate estimates of out-of-sample volatility.

4. **Step 4: select optimal portfolio weights that correspond to $\hat{\lambda}$.** From the solution path obtained in Step 2, select the optimal portfolio weights $\hat{w}$ that correspond to $\hat{\lambda}$ obtained in Step 3.

D.3 ROBUST PORTFOLIO CHOICE

This section briefly describes the key ideas and the implementation steps of the robust-mean-variance approach from the 2021 paper *Robust Portfolio Choice* by Valentina Raponi, Raman Uppal, and Paolo Zaffaroni.[7] The robust-mean-variance approach combined two inefficient "alpha" and "beta" portfolios that collectively produced an efficient mean-variance portfolio. This approach avoided the daunting task of dealing with the model misspecification directly in the mean-variance portfolio and instead leveraged two different techniques to model misspecification applied to the alpha and beta portfolios separately. The beta portfolio could be replaced by $1/N$ without any loss of performance. The misspecification in the alpha portfolio was mitigated using the robust control method described in the 2007 book *Robustness* by Lars Hansen and Thomas Sargent.[8] The authors used the pseudo-Gaussian maximum-likelihood constrained (MLC) estimation approach.

The implementation steps of the robust-mean-variance (RMV) approach are outlined below:

1. **Step 1**. Estimate the parameters of the factor model conditional on the factor realizations without imposing the APT constraints.
2. **Step 2**. Analyze the possibility of pervasive missing factors using PCA to estimate the number of latent pervasive factors.
3. **Step 3**. Estimate the model assuming either small pricing errors or large pricing errors while imposing the APT restrictions.

D.4 BAYESIAN RISK PARITY

This section briefly describes the key formula and idea for applying the framework from the 2017 study *Black-Litterman, Exotic Beta and Varying Efficient Portfolios: An Integrated Approach* by Ricky Cooper and Marat Molyboga for hedge fund portfolios.[9]

We consider a portfolio of N hedge funds with the covariance matrix Σ. We use the equal risk contribution portfolio with weights w_{ERC} as the "prior" portfolio, but it can be replaced with any other portfolio, which is a reasonable candidate for an optimal portfolio, such as $MHRP$. As in Section 5.1.3.2, the reverse optimization step produces the implied expected returns of the hedge funds given risk-parity weights:

$$\pi_{ERC} = \gamma \Sigma w_{ERC}. \tag{D.8}$$

We consider two common opinions regarding hedge fund performance:

- **Equal Sharpe ratio**: all hedge funds are the same and their true Sharpe ratios are equal.
- **Risk-adjusted momentum**: there is performance persistence among hedge funds and past winners have higher ex-ante Sharpe ratios than losers.

As discussed in Section 5.1.3.2, the Black-Litterman framework incorporates investor views that are expressed as statements regarding expected returns of assets with some degree of uncertainty as

$$P\pi \sim N(\nu, \Omega) \tag{D.9}$$

where P is the matrix of opinions, ν is the vector of opinions and Ω is the matrix of uncertainty regarding the opinions.

The **Equal Sharpe** opinion implies that $\frac{\pi_1}{\sigma_1} = \frac{\pi_2}{\sigma_2} = ... = \frac{\pi_N}{\sigma_N}$, which can be captured through $N-1$ opinions: $\frac{\pi_1}{\sigma_1} = \frac{\pi_2}{\sigma_2}, \frac{\pi_2}{\sigma_2} = \frac{\pi_3}{\sigma_3}$, etc.

Thus, the Equal Sharpe opinion can be expressed using the $(N-1)$-by-N matrix P_{ES}

$$P_{ES} = \begin{bmatrix} 1/\sigma_1 & -1/\sigma_2 & 0 & 0 & \dots & 0 & 0 \\ 0 & 1/\sigma_2 & -1/\sigma_3 & 0 & \dots & 0 & 0 \\ & & \dots & & & & \\ 0 & 0 & 0 & 0 & \dots & 1/\sigma_{N-1} & -1/\sigma_N \end{bmatrix} \tag{D.10}$$

and ν_{ES}, an $(N-1)$-by-1 vector of zeros.

Cooper and Molyboga expressed the covariance matrix of the opinion

$$\Omega_{ES} = \frac{1}{c_{ES}} P_{ES} \Sigma P'_{ES}, \tag{D.11}$$

where c_{ES} represented the strength of conviction in the Equal Sharpe opinion with $c_{ES} = 0$ implying having no confidence and $c_{ES} = \infty$ implying certainty.

The final step of the Black-Litterman framework produces the posterior vector of expected excess returns by combining the expectation implied by the "prior" risk-parity portfolio and the expert opinions as two distinct sources of information. Using expression (D.11), we get the following expression for the vector of expected excess returns:

$$\begin{aligned} \pi_{ES} = \pi_{ERC} + \tau \Sigma P'_{ES} \left(P_{ES} \tau \Sigma P'_{ES} + \Omega_{ES} \right)^{-1} \left(\nu_{ES} - P_{ES} \pi_{ERC} \right) = \\ \pi_{ERC} + \tau \Sigma P'_{ES} \left(P_{ES} \tau \Sigma P'_{ES} + \frac{1}{c_{ES}} P_{ES} \Sigma P'_{ES} \right)^{-1} \left(\nu_{ES} - P_{ES} \pi_{ERC} \right) = \\ \pi_{ERC} + \frac{\tau}{\tau + 1/c_{ES}} \Sigma P'_{ES} \left(P_{ES} \Sigma P'_{ES} \right)^{-1} \left(\nu_{ES} - P_{ES} \pi_{ERC} \right). \end{aligned} \tag{D.12}$$

Since ν_{ES} is a vector of zeros, the posterior vector of expected returns that incorporates Equal Sharpe opinion simplifies to:

$$\pi_{ES} = \pi_{ERC} - \frac{\tau}{\tau + 1/c_{ES}} \Sigma P'_{ES} \left(P_{ES} \Sigma P'_{ES} \right)^{-1} P_{ES} \pi_{ERC}. \tag{D.13}$$

Setting τ to $1/T$ produces

$$\pi_{ES} = \pi_{ERC} - \frac{1/T}{1/T + 1/c_{ES}} \Sigma P'_{ES} \left(P_{ES} \Sigma P'_{ES} \right)^{-1} P_{ES} \pi_{ERC}. \tag{D.14}$$

This representation provides guidance for choosing the confidence parameter c_{ES} as a "number of data points" that the opinion represents relative to T, the number of historical observations.

The posterior asset allocation that incorporates the Equal Sharpe view can be calculated as

$$w_{ES} = \frac{1}{\gamma}\Sigma^{-1}\pi_{ES}. \tag{D.15}$$

The **Risk-adjusted momentum** opinion implies that the differences in the future Sharpe ratios are positively related to the differences in past Sharpe ratios (noted with tildes), which could be expressed through $N-1$ opinions: $\frac{\pi_1}{\sigma_1} - \frac{\pi_2}{\sigma_2} = \rho\left(\frac{\tilde{\pi}_1}{\sigma_1} - \frac{\tilde{\pi}_2}{\sigma_2}\right)$, the second row represents the opinion $\frac{\pi_2}{\sigma_2} - \frac{\pi_3}{\sigma_3} = \rho\left(\frac{\tilde{\pi}_2}{\sigma_2} - \frac{\tilde{\pi}_3}{\sigma_3}\right)$, etc. The parameter ρ with a value between 0 and 1 indicates the degree of persistence. $\rho = 0$ implies no performance persistence, which is consistent with the Equal Sharpe view. $\rho = 1$ implies a very high degree of persistence with the expected relative performance equal to the historical relative performance. If ρ is equal to 0.5 that means that if the fund A has historically delivered a Sharpe ratio that is 0.4 higher than that of the fund B (i.e., $Sharpe(A) - Sharpe(B) = 0.4$), then the fund A is expected to deliver the Sharpe ratio, which is $0.4 * 0.5 = 0.2$ higher than that of the fund B.

Thus, the Risk-adjusted momentum opinion can be expressed using the $(N-1)$-by-N matrix P_{RAM}, which is the same as the opinions matrix P_{ES} shown in expression (D.10), but the $(N-1)$-by-1 vector ν_{RAM} is equal to

$$\nu_{RAM} = \rho \begin{bmatrix} \tilde{\pi}_1/\sigma_1 - \tilde{\pi}_2/\sigma_2 \\ \tilde{\pi}_2/\sigma_2 - \tilde{\pi}_3/\sigma_3 \\ \dots \\ \tilde{\pi}_{N-1}/\sigma_{N-1} - \tilde{\pi}_N/\sigma_N \end{bmatrix}. \tag{D.16}$$

The covariance matrix of the opinion

$$\Omega_{RAM} = \frac{1}{c_{RAM}} P_{RAM}\Sigma P'_{RAM}, \tag{D.17}$$

where c_{RAM} represents the strength of conviction in the risk-adjusted momentum.

Using expression (D.12), the vector of expected excess returns that reflects the risk-adjusted momentum view can be expressed as:

$$\pi_{RAM} = \pi_{ERC} + \\ + \frac{\tau}{\tau + 1/c_{RAM}} \Sigma P'_{RAM} \left(P_{RAM}\Sigma P'_{RAM}\right)^{-1} \left(\nu_{RAM} - P_{RAM}\pi_{ERC}\right). \tag{D.18}$$

Setting τ to $1/T$ produces

$$\pi_{RAM} = \pi_{ERC} + \\ + \frac{1/T}{1/T + 1/c_{RAM}} \Sigma P'_{RAM} \left(P_{RAM} \Sigma P'_{RAM} \right)^{-1} \left(\nu_{RAM} - P_{RAM} \pi_{ERC} \right). \tag{D.19}$$

As discussed previously, this representation provides guidance for choosing the confidence parameter c_{RAM} as a "number of data points" that the opinion represents relative to T, the number of historical observations, and the posterior asset allocation that incorporates the risk-adjusted momentum view can be calculated as

$$w_{RAM} = \frac{1}{\gamma} \Sigma^{-1} \pi_{RAM}. \tag{D.20}$$

Glossary

Active management: The attempt to uncover securities the market has either undervalued or overvalued and/or the attempt to time investment decisions to be more heavily invested when the market is rising and less so when the market is falling.

Alpha: A measure of risk-adjusted performance relative to a benchmark. Positive alpha represents outperformance; negative alpha represents underperformance. Positive or negative alpha may be caused by luck, manager skill, costs, and/or wrong choice of benchmark.

Anomaly: Security returns that are not explained by risk considerations per the efficient market hypothesis (EMH).

Arbitrage: The process by which investors attempt to exploit the price difference between two exactly alike (or very similar) securities by simultaneously buying one at a lower price and selling the other at a higher price (thereby avoiding or minimizing risk). The trading activity of arbitrageurs eventually eliminates these price differences.

Asset allocation: The process of determining what percentage of assets should be dedicated to specific asset classes. Also, the end result of this process.

Asset class: A group of assets with similar risk and expected return characteristics. Cash, debt instruments, real estate, and equities are examples of asset classes. Within a major asset class, such as equities, there are more specific classes, such as large- and small-cap company stocks and domestic and international stocks.

Basis point: One one-hundredth of 1 percent, or 0.0001.

Bayes rule (Bayes theorem): A statistical method designed for updating beliefs based on the arrival of new information.

Bayes theorem: See Bayes rule.

Benchmark: An appropriate standard against which hedge funds, mutual funds, and other investment vehicles can be judged. HFR indices are often used to benchmark hedge fund strategies.

Beta: The exposure of a stock, hedge fund, or portfolio to a factor.

Boosting: A popular machine learning method that relies on sequential training of multiple models to improve prediction accuracy. It transforms a system of weak learners (models with low prediction accuracy) into a single strong learner (model with high prediction accuracy).

Calmar ratio: A measure of the return earned above the rate of return on a riskless asset, usually taken as one-month U.S. Treasury bills, relative to the amount of risk taken, with risk being measured by the maximum historical drawdown. For example: The average return earned on an asset was 10 percent. The average rate of a one-month U.S. Treasury bill was 4 percent. The maximum drawdown was 30 percent. The Calmar ratio would be equal to 10 percent minus 4 percent (6 percent) divided by 30 percent, or 0.2.

Capital asset pricing model (CAPM): The first formal asset pricing model. It uses a single factor (market beta) that describes the relationship between risk and expected return, and is used in the pricing of risky securities.

Certainty equivalent return: A certain return that an investor would accept today in exchange for a larger but uncertain future expected return.

Credit default swap (CDS): A financial swap agreement in which the seller of the CDS will compensate the buyer (usually the creditor of the reference loan) in the event of a loan default (by the debtor) or other credit event. In effect, the seller of the CDS insures the buyer against some reference loan defaulting. The buyer of the CDS makes a series of payments (the CDS "fee" or "spread") to the seller and, in exchange, receives a payoff if the loan defaults.

CRSP: The Center for Research in Security Prices is a financial research group at the University of Chicago Business School. The CRSP deciles refer to groups of U.S. stocks ranked by their market capitalization, with CRSP 1 being the largest and CRSP 10 the smallest.

Data mining: A technique for attempting to build predictive real-world models by discerning patterns in masses of historical data.

Diversification: Dividing investment funds among a variety of investments with different risk–return characteristics to minimize portfolio risk.

EAFE Index: The Europe, Australasia, and Far East Index, which consists of the stocks of companies from the developed EAFE countries. Very much like the S&P 500 Index, the stocks within the EAFE index are weighted by market capitalization.

Efficient market hypothesis (EMH): A theory that, at any given time and in a liquid market, security prices fully reflect all available information. The EMH contends that because markets are efficient and current prices reflect all information, attempts to outperform the market are essentially a game of chance rather than one of skill.

Eigenvalue (characteristic value): is a special number associated with a matrix. A correlation matrix with N assets has N eigenvalues. The largest eigenvalue represents the most important common factor, the second largest eigenvalue represents the second most important common factor, which is uncorrelated with the first factor, etc.

Elastic net: A regression approach introduced by Hui Zou and Trevor Hastie. Elastic net is used in statistics and machine learning for variable (factor) selection and prediction. It relies on an L1 norm and an L2 norm penalty functions to eliminate variables with spurious or weak relationships to the dependent variable and improve prediction accuracy.

Emerging markets: The capital markets of less-developed countries that are beginning to acquire characteristics of developed countries, such as higher per-capita income. Countries typically included in this category would be Brazil, Mexico, Romania, Turkey, and Thailand.

Empirical Bayesian approach: A statistical method that uses the data to estimate the prior probability distribution.

Estimation error: The difference between the true value of a parameter and its estimated value.

EMH: See efficient market hypothesis.

Ex-ante: Before the fact.

Excess kurtosis: Excess kurtosis equals Kurtosis - 3. Since normal distribution has kurtosis that is equal to 0, its excess kurtosis is equal to 0. See kurtosis.

Ex-post: After the fact.

EWMA: See exponentially-weighted moving average.

Exponentially-weighted moving average (EWMA): A weighting technique designed to give lower weights to older observations. It is often use for volatility modeling.

Factor: A numerical characteristic or set of characteristics common across a broad set of securities.

False discovery rate: The expected proportion of false discoveries when performing multiple tests simultaneously. See multiple hypothesis testing.

Fama-French four-factor model: Differences in performance between diversified equity portfolios are best explained by the amount of exposure to four factors: the risk of the overall stock market, company size (market capitalization), value (book-to-market ratio), and momentum. Research has shown that, on average, the four factors explain approximately 95 percent of the variation in performance of diversified U.S. stock portfolios.

Fama-French three-factor model: Differences in performance between diversified equity portfolios, which are best explained by three factors: the amount of exposure to the risk of the overall stock market, company size (market capitalization), and value (book-to-market ratio) characteristics. Research has shown that, on average, the three factors explain more than 90 percent of the variation in performance of diversified U.S. stock portfolios.

Funding ratio: A ratio between available assets and liabilities. It is often used to measure the health of defined benefit pension plans. If a funding ratio of a pension fund exceeds one, that indicates that the fund has sufficient assets to cover future liabilities. If a funding ratio is equal to 50 percent, that indicates that the fund is expected to cover 50 percent of future liabilities.

Fung-Hsieh alpha: An regression alpha with respect to the Fung-Hsieh factors. The Fung-Hsieh seven-factor model includes equity market factor (monthly returns of the S&P 500 total return index), size spread factor (monthly returns of the Russell 2000 total return index minus monthly returns of the S&P 500 total return index), bond market factor (the monthly change in the 10-year Treasure constant maturity yield), credit spread factor (the monthly change in the Moody's Baa yield minus 10 year Treasury constant maturity yield), bond trend following factor, currency trend following factor, and commodity trend following factor.

GARCH: See generalized autoregressive conditional heteroskedasticity.

Generalized AutoRegressive Conditional Heteroskedasticity (GARCH): A popular statistical model used for volatility modeling.

Generalized Method of Moments (GMM): A statistical method for estimating parameters without having to fully characterize distributional assumptions. GMM approach is highly relevant for analyzing hedge fund returns because they are non-normal and exhibit positive serial correlation.

Graph theory: A branch of math that studies mathematical structures with pair-wise relations between objects. The objects are called vertices, and their relations are called edges or links.

Heteroscedasticity or heteroskedasticity: Variable volatility. If returns are heteroskedastic, that means that their volatility is not constant.

Hierarchical tree structure: A visual method that displays how a hierarchical system is structured. An organizational chart is an example of a hierarchical tree.

Histogram: A statistical chart introduced by Karl Pearson. It approximates the distribution of a variable by creating bins (or ranges) of values and then counting how many values of the variable fall within each bin.

Homoscedasticity or homoskedasticity: Constant volatility. If returns are homoskedastic, that means that their volatility is constant.

Incentive fee (performance fee): A fee charged by a hedge fund manager based on the fund's performance over the performance period such as a month, a quarter, or a year.

Kurtosis: The degree to which exceptional values, much larger or smaller than the average, occur more frequently (high kurtosis) or less frequently (low kurtosis) than in a normal (bell-shaped) distribution. High kurtosis results in exceptional values called "fat tails." Low kurtosis results in "thin tails."

L1 norm: L1 norm of a vector is equal to the sum of the absolute vector values. It is also known as Manhattan distance.

L2 norm: L2 norm of a vector is equal to the square root of the sum of the squared vector values. It is also known as Euclidean distance.

LASSO: See least absolute shrinkage and selection operator.

Least absolute shrinkage and selection operator (LASSO): A regression approach introduced by Robert Tibshirani. LASSO is used in statistics and machine learning for variable (factor) selection and prediction. It relies on an L1 norm penalty function to eliminate variables with spurious or weak relationships to the dependent variable and improve prediction accuracy.

Ledoit-Wolf shrinkage: A shrinkage estimator of covariance matrices introduced by Olivier Ledoit and Michael Wolf. Ledoit and Wolf proposed several version of the estimator.

Leverage: The use of debt to increase the amount of assets that can be acquired (for example, to buy stock). Leverage increases the riskiness as well as the expected return of a portfolio.

Linear regression: A statistical approach that assumes a linear relationship between independent and dependent variables.

Liquidity: A measure of the ease of trading a security in a market.

Machine learning: A branch of artificial intelligence that studies methods that learn from the data to accomplish a task. Machine learning models fall into three categories: supervised learning, unsupervised learning, and reinforcement learning.

Management fees: Total amount charged to an account for the management of a portfolio.

Market beta: The sensitivity of the return of a stock, mutual fund, or portfolio relative to the return of the overall equity market. Because this was the original form of beta, some refer to market beta as just "beta."

Market cap/market capitalization: For an individual stock, this is the total number of shares of common stock outstanding multiplied by the current price per share. For example, if a company has 100 million shares outstanding and its current stock price is $30 per share, the market cap of the company is $3 billion.

Maximum drawdown: The largest peak-to-valley loss of an investment.

Modern portfolio theory (MPT): A body of academic work founded on four concepts. First, markets are too efficient to allow expected returns in excess of the market's overall rate of return to be achieved consistently through trading systems. Active management is therefore counterproductive. Second, over sustained periods, asset classes can be expected to achieve returns commensurate with their level of risk. Riskier asset classes, such as small companies and value companies, will produce higher returns as compensation for their higher risk. Third, diversification across asset classes can increase returns and reduce risk. For any given level of risk, a portfolio can be constructed to produce the highest expected return. Fourth, there is no one right portfolio for every investor. Each investor must choose an asset allocation that results in a portfolio with an acceptable level of risk for that investor's specific situation.

Monotonic: Changing in such a way that is either never increasing or never decreasing.

MPT: See modern portfolio theory.

MSCI EAFE Index: See EAFE Index.

Multicollinearity issue: A problem that arises in regression analysis when some independent variables are highly correlated. It leads to regressions coefficients that are unstable and difficult to interpret.

Multiple hypothesis testing (multiple comparison problem): A problem that arises when many individual hypothesis texts are performed simultaneously. The bigger the number of tests, the higher the likelihood of false discoveries. Therefore, special statistical techniques are applied to increase the significance threshold for individual discoveries to compensate for the number of tests.

Natural language processing (NLP): A branch of artificial intelligence that teaches computers to process and analyze text data such as documents, articles, or announcements to extract useful information.

Newey-West adjustment (correction): A statistical estimator used in regression analysis to account for serial correlation and heteroskedasticity of residuals. Although it produces the same values of regressions coefficients, the corresponding t-statistics are typically lower.

NFT: See non-fungible tokens.

NLP: See natural language processing.

Non-fungible tokens (NFTs): A unique asset on a blockchain such as digital artwork, domain name, or concert ticket.

Non-synchronous data: Data are synchronous if their components are observed or recorded simultaneously. Otherwise, the data are non-synchronous. For example, daily closing prices of U.S. and European stocks are non-synchronous because the U.S. and European markets close at different times. If an important event happens after the close of the U.S. markets but before the close of the European markets, it will have an impact on the closing prices of the European stocks and will be reflected in the opening prices of the U.S. stocks the following business day.

OLS: See ordinary least squares.

Ordinary least squares (OLS): A standard statistical approach that estimates parameters to minimize the sum of the squares of the differences between the observed dependent variables and their estimated values.

Overfitting: A concept in statistics and machine learning that describes a situation when a model performs very well on training data and poorly on new data. This problem happens when models learn from the noise in the data.

Parsimonious theory: A principle according to which an explanation of a thing or event is made with the fewest possible assumptions.

Performance fee: See incentive fee.

P-value: A statistical measure that is equal to the probability of observing a certain outcome assuming that the null hypothesis is true. For example, if we have a coin and the null hypothesis is that the probability of observing tails and heads is equal to 50 percent, then the p-value of observing two heads in a row is equal to 0.25 ($0.5 * 0.5 = 0.5^2 = 0.25$). However, the p-value of observing 10 heads in a row is approximately equal to 0.1 percent (0.5^{10}).

Random forest (random decision forest): A popular approach in machine learning introduced by Tin Kam Ho. This approach creates a large collection of decision trees that are randomly restricted to the subset of features (variables). Then the output of individual trees is aggregated. Random forests are highly regarded in machine learning for their simplicity and robustness.

Random matrix theory: A branch of math that studies statistical properties of eigenvalues of random matrices.

Rebalancing: The process of restoring a portfolio toward its original asset allocation. Rebalancing can be accomplished either through adding newly investable funds or by selling portions of the best performing asset classes and using the proceeds to purchase additional amounts of the underperforming asset classes.

Risk premium: The higher expected (not guaranteed) return for accepting a specific type of non-diversifiable risk.

Rotationally invariant estimator: A class of methods for estimating eigenvalues of correlation matrices.

Rotationally invariant, optimal shrinkage: A statistical method for estimating eigenvalues of correlation matrices.

R-squared: A statistic that represents the percentage of a fund's or security's movements that can be explained by movements in a benchmark index or set of factors.

Russell 2000 Index: The smallest 2,000 of the largest 3,000 publicly traded U.S. stocks; a common benchmark for small-cap stocks.

Securities and Exchange Commission (SEC): A government agency created by Congress to regulate the securities markets and protect investors. The SEC has jurisdiction over the operation of broker-dealers, investment advisors, mutual funds, and companies selling stocks and bonds to the investing public.

Serial correlation (autocorrelation): A correlation between a variable and its lagged (or delayed) version. Hedge fund returns typically exhibit positive serial correlation. Above average returns tend to be followed by above average returns and below average returns tend to be followed by below average returns.

Sharpe ratio: A measure of the return earned above the rate of return on a riskless asset, usually taken as one-month U.S. Treasury bills, relative to the amount of risk taken, with risk being measured by the standard deviation of returns. For example: The average return earned on an asset was 10 percent. The average rate of a one-month U.S. Treasury bill was 4 percent. The standard deviation was 20 percent. The Sharpe ratio would be equal to 10 percent minus 4 percent (6 percent) divided by 20 percent, or 0.3.

Skewness: A measure of the asymmetry of a distribution. Negative skewness occurs when the values to the left of (less than) the mean are fewer but farther from it than values to the right of (more than) the mean. For example: The return series of –30 percent, 5 percent, 10 percent, and 15 percent has a mean of 0 percent. There is only one return less than 0 percent, and three higher; but the negative one is much farther from zero than the positive ones. Positive skewness occurs when the values to the right of (more than) the mean are fewer but farther from it than values to the left of (less than) the mean.

Small (small-cap) stocks: Small-cap stocks are those of companies considered small relative to other companies, as measured by their market capitalization. Precisely what is considered a "small" company varies by source. For example, one investment professional may define it as having a market capitalization of less than $2 billion, while another may use $5 billion. We are interested in a stock's capitalization because academic evidence indicates that investors can expect to be rewarded by investing in smaller companies' stocks. They are considered to be riskier investments than larger companies' stocks, so investors demand a "risk premium" to invest in them.

S&P 500 Index: A market-cap weighted index of 500 of the largest U.S. stocks, designed to cover a broad and representative sampling of industries.

Sparse regression approach: A class of statistical methods that are efficient at excluding redundant independent variables.

Standard deviation: A measure of volatility or risk. The greater the standard deviation, the greater the volatility of a security or portfolio. Standard deviation can be measured for varying time periods, such as monthly, quarterly, or annually.

Stochastic dominance: A partial order between random variables used extensively in decision theory. The stochastic dominance of order two is particularly relevant in finance because it implies a clear choice for any rational investor. For example, if portfolio A has second-order stochastic dominance over portfolio B, then any rational investor, regardless of the utility function or risk-aversion, would choose A over B.

Systematic risk: Risk that cannot be diversified away. The market must reward investors for assuming systematic risk or they would not take it. That reward is in the form of a risk premium, a higher expected return than could be earned by investing in a less risky instrument.

Tree (decision tree): A popular approach in machine learning that uses a tree-like model of decisions and their outcomes.

T-stat: Short for t-statistic, it is a measure of statistical significance. A value greater than about 2.0 is generally considered meaningfully different from random noise, with a higher number indicating even greater confidence.

Turnover: The trading activity of a fund as it sells securities and replaces them with new ones.

Value stocks: The stocks of companies with relatively low price-to-earnings (P/E) ratios or relatively high book-to-market (BtM) ratios—the opposite of growth stocks. The market anticipates slower earnings growth relative to the overall market. They are considered to be riskier investments than growth companies' stocks, so investors demand a "risk premium" to invest in them.

Volatility: The standard deviation of the change in value of a financial instrument within a specific time horizon. It is often used to quantify the risk of the instrument over that time period. Volatility is typically expressed in annualized terms.

Notes

Chapter 1. Introduction to Hedge Funds

1. "Fashion in Forecasting" by Alfred Winslow Jones, Fortune Magazine, March 1949.

2. "The Jones Nobody Keeps Up With" by Carol Loomis, Fortune Magazine, 1966.

3. Greenlight's white paper: "An Analysis of Allied Capital: Questions of Valuation Techniques."

4. The SEC Release No. 55931: https://www.sec.gov/litigation/admin/2007/34-55931.pdf.

5. "Billionaire Bill Ackman Dumps Herbalife, Ending 5-year War Betting Against It," Investopedia, June 25, 2019.

6. https://www.justice.gov/opa/pr/herbalife-nutrition-ltd-agrees-pay-over-122-million-resolve-fcpa-case.

7. Ellis, C.D. (2014). "The Rise and Fall of Performance Investing." *Financial Analysts Journal,* 70(4): 14–23.

8. Sebastian, M. and S. Attaluri (2014). "Conviction in Equity Investing." *Journal of Portfolio Management,* 40(4): 77–88.

9. Fama, E.F. and K.R. French. (2010). "Luck versus Skill in the Cross-Section of Mutual Fund Returns." *Journal of Finance,* 65(5): 1915–1947.

10. Lee, D.R. and J.A. Verbrugge (1996). "The Efficient Market Theory Thrives on Criticism." *Journal of Applied Corporate Finance,* 9(1): 35–41.

11. Rex Sinquefield, Financial Advisor, March 2001.

12. "Will Nepo's Supercomputer Give Him World Chess Title Edge over Carlsen?", The Guardian, Nov 25, 2021.

13. Amin, G.S. and H.M. Kat (2003). "Hedge Fund Performance 1990–2000: Do the 'Money Machines' Really Add Value?" *Journal of Financial and Quantitative Analysis,* 38(2): 251–274.

14. Kosowski, R., Naik, N.Y., and M. Teo (2007). "Do Hedge Funds Deliver Alpha? A Bayesian and Bootstrap Analysis." *Journal of Financial Economics,* 84(1): 229–264.

15. Barth, D., Joenvaara, J., Kauppila, M., and R. Wermers (2021). "The Hedge Fund Industry is Bigger (and has Performed Better) Than You Think." *SSRN working paper.*

16. Berk, J.B. and R.C. Green (2004). "Mutual Fund Flows and Performance in Rational Markets." *Journal of Political Economy,* 112(6): 1269–1295.

17. Baquero, G. and M. Verbeek. (2022). "Hedge Fund Flows and Performance Streaks: How Investors Weight Information" *Management Science,* 68: 1–19.

18. Agarwal, V., Green, T.C., and R. Honglin (2018). "Alpha or Beta in the Eye of the Beholder: What Drives Hedge Fund Flows?" *Journal of Financial Economics,* 127(3): 417–434.

19. Getmansky M. (2012). "The Life Cycle of Hedge Funds: Fund Flows, Size, Competition, and Performance." *Quarterly Journal of Finance,* 2(1): 1–53.

20. Weidenmuller, O. and M. Verbeeky (2009). "Crowded Chickens Farm Fewer Eggs: Capacity Constraints in the Hedge Fund Industry Revisited." *SSRN working paper.*

21. Bernstein, W. (2012). "Skating Where the Puck Was." *Efficient Frontier Publications.*

22. Bollen, N., Joenvaara, J., and M. Kauppila (2021). "Hedge Fund Performance: End of an Era?" *Financial Analysts Journal,* 77(3): 109–132.

23. Agarwal, V., Aragon, G., Nanda, V., and K. Wei (2022). "Anticipatory Trading Against Distressed Mega Hedge Funds." *SSRN working paper.*

24. Asness, C.S., Krail, R.J., and J.M. Liew (2001). "Do Hedge Funds Hedge?" *Journal of Portfolio Management,* 28(1): 6–19.

25. Getmansky, M., Lo, A.W., and I. Makarov (2004). "An Econometric Model of Serial Correlation and Illiquidity in Hedge Fund Returns." *Journal of Financial Economics,* 74(3): 529–609.

26. Dimson, E. (1979). "Risk Measurement When Shares are Subject to Infrequent Trading." *Journal of Financial Economics,* 7(2): 197–226.

27. Scholes, M. and J.T. Williams (1977). "Estimating Betas from Nonsynchronous Data." *Journal of Financial Economics,* 5(3): 309–327.

28. Billio, M., Frattarolo, L., and L. Pelizzon (2016). "Hedge Fund Tail Risk: an Investigation in Stressed Markets." *Journal of Alternative Investments,* 18(4): 109–124.

29. Kazemi, M. and E. Islamaj (2016). "Returns to Active Management: The Case of Hedge Funds." *SSRN working paper.*

30. Liang, H., Sun, L., and M. Teo (2021). "Responsible Hedge Funds." *SSRN working paper.*

31. Kim, S. and A. Yoon (2021). "Analyzing Active Fund Managers' Commitment to ESG: Evidence from the United Nations Principles for Responsible Investing." *SSRN working paper.*

32. Gibson, R., Glossner, S., Krueger, P., Matos, P., and T. Steffen (2022). "Do Responsible Investors Invest Responsibly?" *SSRN working paper.*

33. Cao, C., Chen, Y., Goetzmann, W.N., and B. Liang (2018). "Hedge Funds and Stock Price Formation." *Financial Analysts Journal,* 74(3): 54–68.

Chapter 2. Hedge Fund Research and Data

1. Welch, I. (2013). "A Critique of Recent Quantitative and Deep-structure Modeling in Capital Structure Research and Beyond." *Critical Finance Review,* 2(1): 131–172.

2. Fama, E.F. and K.R. French (1992). "The Cross-section of Expected Stock Returns." *Journal of Finance,* 47(2): 427–465.

3. Jorion, P. and C. Schwarz (2014). "The Strategic Listing Decisions of Hedge Funds." *Journal of Financial and Quantitative Analysis,* 49(3): 773–796.

4. Bollen, N.P. and V.K. Pool (2009). "Do Hedge Fund Managers Misreport Returns? Evidence from the Pooled Distribution." *Journal of Finance,* 64(5): 2257–2288.

5. Joenvaara, J., Kauppila, M., Kosowski, R., and P. Tolonen (2021). "Hedge Fund Performance: Are Stylized Facts Sensitive to Which Database One Uses?" *Critical Finance Review,* 10(2): 271–327.

6. Fung, W. and D.A. Hsieh (2002). "Hedge Fund Benchmarks: Information Content and Biases." *Financial Analysts Journal,* 58(1): 22–34.

7. Aggarwal, R.K. and P. Jorion (2010). "Hidden Survivorship Bias in Hedge Fund Returns." *Financial Analysts Journal,* 66(2): 69–74.

8. Fung, W. and D.A. Hsieh (2009). "Measurement Biases in Hedge Fund Performance Data: An Update." *Financial Analysts Journal*, 65(3): 36–38.

9. Aiken, A.L., Clifford, C.P., and J. Ellis (2013). "Out of the Dark: Hedge Fund Reporting Biases and Commercial Databases." *Review of Financial Studies*, 26(1): 208–243.

10. Edelman, D., Fung, W., and D.A. Hsieh (2013). "Exploring Unchartered Territories of the Hedge Fund Industry: Empirical Characteristics of Mega Hedge Fund Returns." *Journal of Financial Economics*, 109(3): 734–758.

11. Jorion, P. and C. Schwarz (2014). "The Strategic Listing Decisions of Hedge Funds." *Journal of Financial and Quantitative Analysis*, 49(3): 773–796.

12. Ackermann, C., McEnally, R., and D. Ravenscraft (2014). "The Performance of Hedge Funds: Risk, Return, and Incentives." *Journal of Finance*, 54(3): 833–874.

13. Joenvaara, J., Kauppila, M., Kosowski, R., and P. Tolonen (2021). "Hedge Fund Performance: Are Stylized Facts Sensitive to Which Database One Uses?" *Critical Finance Review*, 10(2): 271–327.

14. Barth, D., Joenvaara, J., Kauppila, M., and R. Wermers (2021). "The Hedge Fund Industry is Bigger (and has Performed Better) Than You Think." *SSRN working paper.*

15. Wallis, W.A. (1980). "The Statistical Research Group, 1942-1945." *Journal of the American Statistical Association*, 75(370): 320–330.

16. Liang, B. (2000). "Hedge Funds: the Living and the Dead." *Journal of Financial and Quantitative Analysis*, 35(3): 309–325.

17. Brown, S.J., Goetzmann, W.N., and R.G. Ibbotson (1999). "Offshore Hedge Funds: Survival and Performance 1989-1995." *Journal of Business*, 72(1): 91–118.

18. Brown, S.J., Goetzmann, W.N., Ibbotson, R.G., and S.A. Ross (1992). "Survivorship Bias in Performance Studies." *Review of Financial Studies*, 5(4): 553–580.

19. Malkiel, B.G. and A. Saha (2005). "Hedge Funds: Risk and Return." *Financial Analysts Journal*, 61(6): 80–88.

20. Kosowski, R., Naik, N.Y., and M. Teo (2007). "Do Hedge Funds Deliver Alpha? A Bayesian and Bootstrap Analysis." *Journal of Financial Economics*, 84(1): 229–264.

21. Bhardwaj, G., Gorton, G.B., and K.G. Rouwenhorst (2014). "Fooling Some of the People All of the Time: The Inefficient Performance and Persistence of

Commodity Trading Advisors." *Review of Financial Studies,* 27(11): 3099–3132.

22. Jorion, P. and C. Schwarz (2017). "The Fix is In: Properly Backing Out Backfill Bias." *Review of Financial Studies,* 32(12): 5048–5099.

23. Fung, W. and D.A. Hsieh (2009). "Measurement Biases in Hedge Fund Performance Data: An Update." *Financial Analysts Journal,* 65(3): 36–38.

24. Fama, E.F. and K.R. French (2010). "Luck Versus Skill in the Cross-section of Mutual Fund Returns." *Journal of Finance,* 65(5): 1915–1947.

25. Foran, J., Hutchinson, M.C., McCarthy, D.F., and J. O'Brien (2017). "Just a One Trick Pony? An Analysis of CTA Risk and Return." *Journal of Alternative Investments,* 20(2): 8–26.

26. Jorion, P. and C. Schwarz (2017). "The Fix is In: Properly Backing Out Backfill Bias." *Review of Financial Studies,* 32(12): 5048–5099.

27. Ackermann, C., McEnally, R., and D. Ravenscraft (2014). "The Performance of Hedge Funds: Risk, Return, and Incentives." *Journal of Finance,* 54(3): 833–874.

28. Bhardwaj, G., Gorton, G.B., and K.G. Rouwenhorst (2014). "Fooling Some of the People All of the Time: The Inefficient Performance and Persistence of Commodity Trading Advisors." *Review of Financial Studies,* 27(11): 3099–3132.

29. Patton, A.J., Ramadorai, T., and M. Streatfield (2015). "Change You Can Believe In? Hedge Fund Data Revisions." *Journal of Finance,* 70(3): 963–999.

30. Capocci, D. and G. Hubner (2004). "Analysis of Hedge Fund Performance." *Journal of Empirical Finance,* 11(1): 55–89.

31. Kosowski, R., Naik, N.Y., and M. Teo (2007). "Do Hedge Funds Deliver Alpha? A Bayesian and Bootstrap Analysis." *Journal of Financial Economics,* 84(1): 229–264.

32. Jagannathan, R., Malakhov, A., and D. Novikov (2010). "Do Hot Hands Exist Among Hedge Fund Managers? An Empirical Evaluation." *Journal of Finance,* 65(1): 217–255.

33. Fama, E.F. and K.R. French (1992). "The Cross-section of Expected Stock Returns." *Journal of Finance,* 47(2): 427–465.

34. Molyboga, M. and C. L'Ahelec (2016). "A Simulation-based Methodology for Evaluating Hedge Fund Investments." *Journal of Asset Management,* 17(6): 434–452.

35. Molyboga, M., Baek, S., and J. Bilson (2017). "Assessing Hedge Fund Performance with Institutional Constraints: Evidence from CTA Funds." *Journal of Asset Management,* 18: 547–565.

Chapter 3. Manager Selection and Hedge Fund Factors

1. Asness, C.S., Moskowitz, T.J., and L.H. Pedersen (2013). "Value and Momentum Everywhere." *Journal of Finance,* 68(3): 929–985.

2. Hendricks, D., Patel, J., and R. Zeckhauser (1993). "Hot Hands in Mutual Funds: Short-run Persistence of Relative Performance, 1974-1988." *Journal of Finance,* 48(1): 93–130.

3. Carhart, M. (1997). "On Persistence in Mutual Fund Performance." *Journal of Finance,* 52(1): 57–82.

4. Capocci, D. and G. Hubner (2004). "Analysis of Hedge Fund Performance." *Journal of Empirical Finance,* 11(1): 55–89.

5. Agarwal,V. and Naik, N.Y. (2000). "Multi-Period Performance Persistence Analysis of Hedge Funds." *Journal of Financial and Quantitative Analysis,* 35(3): 327–342.

6. Agarwal,V. and Naik, N.Y. (2000). "On Taking the Alternative Route: The Risks, Rewards, and Performance Persistence of Hedge Funds." *Journal of Alternative Investments,* 2(4): 6–23.

7. Kosowski, R., Naik, N.Y., and M. Teo (2007). "Do Hedge Funds Deliver Alpha? A Bayesian and Bootstrap Analysis." *Journal of Financial Economics,* 84(1): 229–264.

8. Pastor, L. and R.F. Stambaugh (2002). "Mutual Fund Performance and Seemingly Unrelated Assets." *Journal of Financial Economics,* 63(3): 315–349.

9. Jagannathan, R., Malakhov, A., and D. Novikov (2010). "Do Hot Hands Exist Among Hedge Fund Managers? An Empirical Evaluation." *Journal of Finance,* 65(1): 217–255.

10. Molyboga, M. and C. L'Ahelec (2016). "A Simulation-Based Methodology for Evaluating Hedge Fund Investments." *Journal of Asset Management,* 17(6): 434–452.

11. Molyboga, M., Baek, S., and J. Bilson (2017). "Assessing Hedge Fund Performance with Institutional Constraints: Evidence from CTA Funds." *Journal of Asset Management,* 18: 547–565.

12. Capocci, D. and G. Hubner (2004). "Analysis of Hedge Fund Performance." *Journal of Empirical Finance,* 11(1): 55–89.

13. Jagannathan, R., Malakhov, A., and D. Novikov (2010). "Do Hot Hands Exist Among Hedge Fund Managers? An Empirical Evaluation." *Journal of Finance,* 65(1): 217–255.

14. Jagannathan, R., Malakhov, A., and D. Novikov (2010). "Do Hot Hands Exist Among Hedge Fund Managers? An Empirical Evaluation." *Journal of Finance,* 65(1): 217–255.

15. Molyboga, M. and C. L'Ahelec (2016). "A Simulation-Based Methodology for Evaluating Hedge Fund Investments." *Journal of Asset Management,* 17(6): 434–452.

16. Molyboga, M., Baek, S., and J. Bilson (2017). "Assessing Hedge Fund Performance with Institutional Constraints: Evidence from CTA Funds." *Journal of Asset Management,* 18: 547–565.

17. Molyboga, M. and C. L'Ahelec (2016). "A Simulation-Based Methodology for Evaluating Hedge Fund Investments." *Journal of Asset Management,* 17(6): 434–452.

18. Molyboga, M., Baek, S., and J. Bilson (2017). "Assessing Hedge Fund Performance with Institutional Constraints: Evidence from CTA Funds." *Journal of Asset Management,* 18: 547–565.

19. DeMiguel, V., Garlappi, L., and R. Uppal (2009). "Optimal versus Naive Diversification: How Inefficient is the 1/N Portfolio Strategy?" *Review of Financial Studies,* 22(5): 1915–1953.

20. Molyboga, M., Baek, S., and J. Bilson (2017). "Assessing Hedge Fund Performance with Institutional Constraints: Evidence from CTA Funds." *Journal of Asset Management,* 18: 547–565.

21. Joenvaara, J., Kauppila, M., Kosowski, R., and P. Tolonen (2009). "Hedge Fund Performance: Are Stylized Facts Sensitive to Which Database One Uses?" *Critical Finance Review,* 10(2): 271–327.

22. Ackermann, C., McEnally, R., and D. Ravenscraft (2014). "The Performance of Hedge Funds: Risk, Return, and Incentives." *Journal of Finance,* 54(3): 833–874.

23. Bhardwaj, G., Gorton, G.B., and K.G. Rouwenhorst (2014). "Fooling Some of the People All of the Time: The Inefficient Performance and Persistence of Commodity Trading Advisors." *Review of Financial Studies,* 27(11): 3099–3132.

24. Foran, J., Hutchinson, M.C., McCarthy, D.F., and J. O'Brien (2017). "Just a One Trick Pony? An Analysis of CTA Risk and Return." *Journal of Alternative Investments,* 20(2): 8–26.

25. Jorion, P. and C. Schwarz (2017). "The Fix is In: Properly Backing Out Backfill Bias." *Review of Financial Studies,* 32(12): 5048–5099.

26. For example, Robert Kosowski, Narayan Naik, and Melvyn Teo, in their 2007 study "Do Hedge Funds Deliver Alpha? A Bayesian and Bootstrap Analysis" used a fixed AUM level of $20 million.

27. Efron, B. (1979). "Bootstrap Methods: Another Look at the Jackknife." *Annals of Statistics,* 7(1): 1–26.

28. Efron, B. and G. Gong (1983). "A Leisurely Look at the Bootstrap, the Jackknife, and Cross-validation." *American Statistician,* 37(1): 36–48.

29. Molyboga, M., Baek, S., and J. Bilson (2017). "Assessing Hedge Fund Performance with Institutional Constraints: Evidence from CTA Funds." *Journal of Asset Management,* 18: 547–565.

30. Fischmar, D. and C. Peters (1991). "Portfolio Analysis of Stocks, Bonds, and Managed Futures Using Compromise Stochastic Dominance." *Journal of Futures Markets,* 11(3): 259–270.

31. Molyboga, M., Baek, S., and J. Bilson (2017). *Assessing Hedge Fund Performance with Institutional Constraints: Evidence from CTA Funds. Journal of Asset Management,* 18: 547–565.

32. Joenvaara, J., Kauppila, M., Kosowski, R., and P. Tolonen (2009). "Hedge Fund Performance: Are Stylized Facts Sensitive to Which Database One Uses?" *Critical Finance Review,* 10(2): 271–327.

33. Chen, N.F., Roll, R., and S.A. Ross (1986). "Economic Forces and the Stock Market." *Journal of Business,* 59(3): 383–403.

34. Fama, E.F. and K.R. French (1996). "Multifactor Explanations of Asset Pricing Anomalies." *Journal of Finance,* 51(1): 55–84.

35. Fung, W. and D.A. Hsieh (2004). "Hedge Fund Benchmarks: A Risk-based Approach." *Financial Analysts Journal,* 60(5): 65–80.

36. The monthly returns for the Fung-Hsieh factors along with their descriptions are available at the David A. Hsieh data library: https://people.duke.edu/~dah7/HFRFData.htm.

37. Fung, W. and D.A. Hsieh (2001). "The Risk in Hedge Fund Strategies: Theory and Evidence from Trend Followers." *Review of Financial Studies,* 14(2): 313–341.

38. Ibid.

39. Capocci, D. and G. Hubner (2004). "Analysis of Hedge Fund Performance." *Journal of Empirical Finance,* 11(1): 55–89.

40. Kosowski, R., Naik, N.Y., and M. Teo (2007). "Do Hedge Funds Deliver Alpha? A Bayesian and Bootstrap Analysis." *Journal of Financial Economics,* 84(1): 229–264.

41. Jagannathan, R., Malakhov, A., and D. Novikov (2010). "Do Hot Hands Exist Among Hedge Fund Managers? An Empirical Evaluation." *Journal of Finance,* 65(1): 217–255.

42. Joenvaara, J., Kauppila, M., Kosowski, R., and P. Tolonen (2021). "Hedge Fund Performance: Are Stylized Facts Sensitive to Which Database One Uses?" *Critical Finance Review,* 10(2): 271–327.

43. Agarwal, V., Daniel, N.D., and N.Y. Naik (2009). "Role of Managerial Incentives and Discretion in Hedge Fund Performance." *Journal of Finance,* 64(5): 2221–2256.

44. Bollen, N.P. (2013). "Zero-R^2 Hedge Funds and Market Neutrality." *Journal of Financial and Quantitative Analysis,* 48(2): 519–547.

45. Bhardwaj, G., Gorton, G.B., and K.G. Rouwenhorst (2014). "Fooling Some of the People All of the Time: The Inefficient Performance and Persistence of Commodity Trading Advisors." *Review of Financial Studies,* 27(11): 3099–3132.

46. http://people.duke.edu/~dah7/DataLibrary/TF-Fac.xls

47. Moskowitz, T.J., Ooi, Y.H., and L.H. Pedersen (2012). "Time-Series Momentum." *Journal of Financial Economics,* 104(2): 228–250.

48. Joenvaara, J., Kauppila, M., Kosowski, R., and P. Tolonen (2021). "Hedge Fund Performance: Are Stylized Facts Sensitive to Which Database One Uses?" *Critical Finance Review,* 10(2): 271–327.

49. Carhart, M. (1997). "On Persistence in Mutual Fund Performance." *Journal of Finance,* 52(1): 57–82.

50. Fama, E.F. and K.R. French (2012). "Size, Value, and Momentum in International Stock Returns." *Journal of Financial Economics,* 105(3): 457–472.

51. Asness, C.S., Moskowitz, T.J., and L.H. Pedersen (2013). "Value and Momentum Everywhere." *Journal of Finance,* 68(3): 929–985.

52. Moskowitz, T.J., Ooi, Y.H., and L.H. Pedersen (2012). "Time-Series Momentum." *Journal of Financial Economics,* 104(2): 228–250.

53. Pastor, L. and R.F. Stambaugh (2003). "Liquidity Risk and Expected Stock Return." *Journal of Political Economy,* 111(3): 642–685.

54. Frazzini, A. and L.H. Pedersen (2014). "Betting Against Beta." *Journal of Financial Economics,* 111(1): 1–25.

55. Gu, S., Kelly, B., and D. Xiu (2020). "Empirical Asset Pricing via Machine Learning." *Review of Financial Studies,* 33(5): 2223–2273.

56. Tibshirani, R. (1996). "Regression Shrinkage and Selection via the Lasso." *Journal of the Royal Statistical Society: Series B (Statistical Methodology),* 58(1): 267–288.

57. Zou, H. and T. Hastie (2005). "Regularization and Variable Selection via the Elastic Net." *Journal of the Royal Statistical Society: Series B (Statistical Methodology),* 67(2): 301–320.

58. Bollen, N.P. (2013). "Zero-R^2 Hedge Funds and Market Neutrality." *Journal of Financial and Quantitative Analysis,* 48(2): 519–547.

59. Molyboga, M., Swedroe, L., and J. Qian (2020). "Short-Term Trend: A Jewel Hidden in Daily Returns." *Journal of Portfolio Management,* 47(1): 154–167.

60. Moskowitz, T.J., Ooi, Y.H., and L.H. Pedersen (2012). "Time-Series Momentum." *Journal of Financial Economics,* 104(2): 228–250.

61. Hurst, B., Ooi,Y.H., and L.H. Pedersen (2017). "A Century of Evidence on Trend-following Investing." *Journal of Portfolio Management,* 44(1): 15–29.

62. Szymanowska, M., De Roon, F., Nijman, T., and R. Van Den Goorbergh (2014). "An Anatomy of Commodity Futures Risk Premia." *Journal of Finance,* 69(1): 453–482.

63. Blocher, J., Cooper, R., and M. Molyboga (2018). "Benchmarking Commodity Investments." *Journal of Futures Markets,* 38(3): 340–358.

64. Newey, W. and K. West (1987). "A Simple, Positive Semi-definite, Heteroskedasticity and Autocorrelation Consistent Covariance Matrix." *Econometrica,* 55(3): 703–708.

65. Buchanan, B. (2019). "Artificial Intelligence in Finance." *http://doi.org/10.5281/zenodo.2612537.*

66. Gu, S., Kelly, B., and D. Xiu (2020). "Empirical Asset Pricing via Machine Learning." *Review of Financial Studies,* 33(5): 2223–2273.

67. Rapach, D., Strauss, J., and G. Zhou (2013). "International Stock Return Predictability: What is the Role of the United States?" *Journal of Finance,* 68(4): 1633–1662.

68. Rapach, D., Strauss, J., and G. Zhou (2019). "Industry Return Predictability: A Machine Learning Approach." *Journal of Financial Data Science,* 1(3): 9–28.

69. DeMiguel, V., Utrera, A.M., Nogales, F.J., and R. Uppal (2020). "A Transaction-cost Perspective on the Multitude of Firm Characteristics." *Review of Financial Studies,* 33(5): 2180–2222.

70. Feng, G., Giglio, S., and D. Xiu (2020). "Taming the Zoo: A Test of New Factors." *Journal of Finance,* 75(3): 1327–1370.

71. Freyberger, J., Neuhierl, A., and M. Weber (2020). "Dissecting Characteristics Nonparametrically." *Review of Financial Studies,* 33(5): 2326–2377.

72. Mascio, D. and F. Fabozzi (2019). "Sentiment Indices and Their Forecasting Ability." *Journal of Forecasting,* 38(4): 257–276.

73. Mascio, D. and F. Fabozzi (2021). "Market Timing Using Combined Forecasts and Machine Learning." *Journal of Forecasting,* 40(1): 1–16.

74. Zou, H. and T. Hastie (2005). "Regularization and Variable Selection via the Elastic Net." *Journal of the Royal Statistical Society: Series B (Statistical Methodology),* 67(2): 301–320.

75. Gu, S., Kelly, B., and D. Xiu (2020). "Empirical Asset Pricing via Machine Learning." *Review of Financial Studies,* 33(5): 2223–2273.

76. Belloni, A., Chen, D., Chernozhukov, V., and C. Hansen (2012). "Sparse Models and Methods for Optimal Instruments with an Application to Eminent Domain." *Econometrica,* 80(6): 2369–2429.

77. Feng, G., Giglio, S., and D. Xiu (2020). "Taming the Zoo: A Test of New Factors." *Journal of Finance,* 75(3): 1327–1370.

78. Rzepczynski, M. and K. Black (2021). "Alternative Investment Due Diligence: A Survey on Key Drivers for Manager Selection." *CAIA Association.*

79. Scharfman, J. (2022). "Operational Due Diligence on Cryptocurrency and Digital Asset Funds." *Journal of Alternative Investments,* 24(3): 44–50.

80. "Japan Cracks Down on Cryptocurrency Exchanges after $534 Million Heist; Police Begin Investigation," Forbes, January 30, 2018.

81. "The False Narrative of Bitcoin's Role in Illicit Activity", Forbes, January 19, 2021.

Chapter 4. Performance Persistence

1. Kosowski, R., Timmermann, A., Wermers, R., and H. White (2006). "Can Mutual Fund 'Stars' Really Pick Stocks? New Evidence from a Bootstrap Analysis." *Journal of Finance,* 61(6): 2551–2595.

2. Kosowski, R., Naik, N.Y., and M. Teo (2007). "Do Hedge Funds Deliver Alpha? A Bayesian and Bootstrap Analysis." *Journal of Financial Economics,* 84(1): 229–264.

3. Fama, E.F. and K.R. French (2010). "Luck versus Skill in the Cross-Section of Mutual Fund Returns." *Journal of Finance,* 65(5): 1915–1947.

4. Barras, L., Scaillet, O., and R. Wermers (2010). "False Discoveries in Mutual Fund Performance: Measuring Luck in Estimated Alphas." *Journal of Finance,* 65(1): 179–216.

5. Ibid.

6. Ibid.

7. Barras, L., Scaillet, O., and R. Wermers (2010). "False Discoveries in Mutual Fund Performance: Measuring Luck in Estimated Alphas." *Journal of Finance,* 65(1): 179–216.

8. Barras, L., Scaillet, O., and R. Wermers (2010). "False Discoveries in Mutual Fund Performance: Measuring Luck in Estimated Alphas." *Journal of Finance,* 65(1): 179–216.

9. Harvey, C.R. and Y. Liu (2018). "Detecting Repeatable Performance." *Review of Financial Studies,* 31(7): 2499–2552.

10. Kosowski, R., Naik, N.Y., and M. Teo (2007). "Do Hedge Funds Deliver Alpha? A Bayesian and Bootstrap Analysis." *Journal of Financial Economics,* 84(1): 229–264.

11. Barras, L., Scaillet, O., and R. Wermers (2010). "False Discoveries in Mutual Fund Performance: Measuring Luck in Estimated Alphas." *Journal of Finance,* 65(1): 179–216.

12. Pastor, L. and R.F. Stambaugh (2002). "Mutual Fund Performance and Seemingly Unrelated Assets." *Journal of Financial Economics,* 63(3): 315–349.

13. Kosowski, R., Naik, N.Y., and M. Teo (2007). "Do Hedge Funds Deliver Alpha? A Bayesian and Bootstrap Analysis." *Journal of Financial Economics,* 84(1): 229–264.

14. Berk, J.B. and R.C. Green (2004). "Mutual Fund Flows and Performance in Rational Markets." *Journal of Political Economy,* 112(6): 1269–1295.

15. Getmansky M. (2012). "The Life Cycle of Hedge Funds: Fund Flows, Size, Competition, and Performance." *Quarterly Journal of Finance,* 2(1): 1–53.

16. Roussanov, N., Ruan, H., and Y. Wei (2021). "Marketing Mutual Funds." *Review of Financial Studies,* 34(6): 3045–3094.

17. Roussanov, N., Ruan, H., and Y. Wei (2022). "Mutual Fund Flows and Performance in (Imperfectly) Rational Markets?" *SSRN working paper.*

18. Forsberg, D., Gallagher, D. R., and G. J. Warren (2021). "Identifying Hedge Fund Skill by Using Peer Cohorts." *Financial Analysts Journal,* 77(2): 97–123.

19. Bollen, N.P. (2013). "Zero-R^2 Hedge Funds and Market Neutrality." *Journal of Financial and Quantitative Analysis,* 48(2): 519–547.

20. Titman, S. and C. Tiu (2011). "Do the Best Hedge Funds Hedge?" *Review of Financial Studies,* 24(1): 123–168.

21. Fama, E.F. and J.D. MacBeth (1973). "Risk, Return, and Equilibrium: Empirical Tests." *Journal of Political Economy,* 81(3): 607–636.

22. Jagannathan, R., Malakhov, A., and D. Novikov (2010). "Do Hot Hands Exist Among Hedge Fund Managers? An Empirical Evaluation." *Journal of Finance,* 65(1): 217–255.

23. Kosowski, R., Naik, N.Y., and M. Teo (2007). "Do Hedge Funds Deliver Alpha? A Bayesian and Bootstrap Analysis." *Journal of Financial Economics,* 84(1): 229–264.

Chapter 5. From Mean-Variance to Risk Parity

1. Molyboga, M., Baek, S., and J. Bilson (2017). "Assessing Hedge Fund Performance with Institutional Constraints: Evidence from CTA Funds." *Journal of Asset Management,* 18: 547–565.

2. Molyboga, M. and C. L'Ahelec (2016). "A Simulation-Based Methodology for Evaluating Hedge Fund Investments." *Journal of Asset Management,* 17(6): 434–452.

3. DeMiguel, V., Garlappi, L., and R. Uppal (2009). "Optimal versus Naive Diversification: How Inefficient is the 1/N Portfolio Strategy?" *Review of Financial Studies,* 22(5): 1915–1953.

4. Markowitz, H.M. (1952). "Portfolio Selection." *Journal of Finance,* 7(1): 77–91.

5. Michaud, R.O. (1989). "The Markowitz Optimization Enigma: Is Optimized Optimal?" *Financial Analysts Journal,* 45(1): 31–42.

6. Best, M.J. and R.R. Grauer (1991). "On the Sensitivity of Mean-Variance-Efficient Portfolios to Changes in Asset Means: Some Analytical and Computational Results." *Review of Financial Studies,* 4(2): 315–342.

7. In this illustrative example, portfolio weights are optimized to produce the highest Sharpe ratio.

8. Merton, R.C. (1980). "On Estimating the Expected Return on the Markets: An Exploratory Investigation." *Journal of Financial Economics,* 8(4): 323–361.

9. Bailey, D. and M. Lopez de Prado (2012). *Balanced Baskets: A New Approach to Trading and Hedging Risks. Journal of Investment Strategies,* 1(4): 21–62.

10. Yang, Z., Han, K., Molyboga, M., and G. Molyboga (2016). *A New Diagnostic Approach to Evaluating the Stability of Optimal Portfolios. Journal of Investing,* 25(1): 37–45.

11. Merton, R.C. (1980). "On Estimating the Expected Return on the Markets: An Exploratory Investigation." *Journal of Financial Economics,* 8(4): 323–361.

12. DeMiguel, V., Garlappi, L., and R. Uppal (2009). "Optimal versus Naive Diversification: How Inefficient is the 1/N Portfolio Strategy?" *Review of Financial Studies,* 22(5): 1915–1953.

13. Stein, C. (1956). "Inadmissibility of the Usual Estimator for the Mean of a Multivariate Normal Distribution." *Proceedings of the Third Berkeley Symposium on Mathematical Statistics and Probability,* 1: 197–206.

14. Jorion, P. (1986). "Bayes-Stein Estimation for Portfolio Analysis." *Journal of Financial and Quantitative Analysis,* 21(3): 279–292.

15. Ledoit, O. and L. Wolf (2003). "Improved Estimation of the Covariance Matrix of Stock Returns with an Application to Portfolio Selection." *Journal of Empirical Finance,* 10(5): 603–621.

16. Ledoit, O. and L. Wolf (2004). "Honey, I Shrunk the Sample Covariance Matrix." *Journal of Portfolio Management,* 30(4): 110–119.

17. Black, F. and R. Litterman (1992). "Global Portfolio Optimization." *Financial Analysts Journal,* 48(5): 28–43.

18. Sharpe, W.F. (1964). "Capital Asset Prices: A Theory of Market Equilibrium Under Conditions of Risk." *Journal of Finance,* 19(4): 425–442.

19. Green, R.C. and B. Hollifield (1992). "When Will Mean-Variance Efficient Portfolios Be Well Diversified?" *Journal of Finance,* 47(5): 1785–1809.

20. Jagannathan, R. and T. Ma (2003). "Risk Reduction in Large Portfolios: Why Imposing the Wrong Constraints Helps." *Journal of Finance,* 58(4): 1651–1683.

21. DeMiguel, V., Garlappi, L., and R. Uppal (2009). "Optimal versus Naive Diversification: How Inefficient is the 1/N Portfolio Strategy?" *Review of Financial Studies,* 22(5): 1915–1953.

22. Maillard, S., Roncalli, T., and J. Teiletche (2010). "The Properties of Equally Weighted Risk Contribution Portfolios." *Journal of Portfolio Management,* 36(4): 60–70.

23. Ibid.

24. MacKinlay, A.C. and L. Pastor (2000). "Asset Pricing Models: Implications for Expected Returns and Portfolio Selection." *Review of Financial Studies,* 13(4): 883–916.

25. Kan, R. and G. Zhou (2007). "Optimal Portfolio Choice with Parameter Uncertainty." *Journal of Financial and Quantitative Analysis,* 42(3): 621–656.

26. Garlappi, L., Uppal, R., and T. Wang (2007). "Portfolio Selection with Parameter and Model Uncertainty: A Multi-Prior Approach." *Review of Financial Studies,* 20(1): 41–81.

27. Jagannathan, R. and T. Ma (2003). "Risk Reduction in Large Portfolios: Why Imposing the Wrong Constraints Helps." *Journal of Finance,* 58(4): 1651–1683.

28. DeMiguel, V., Garlappi, L., and R. Uppal (2009). "Optimal versus Naive Diversification: How Inefficient is the 1/N Portfolio Strategy?" *Review of Financial Studies,* 22(5): 1915–1953.

29. Kritzman, M., Page, S., and D. Turkington (2010). "In Defense of Optimization: The Fallacy of $1/N$." *Financial Analysts Journal,* 66(2): 31–39.

30. Hallerbach, W. (2010). "A Proof of Optimality of Volatility Weighting over Time." *Journal of Investment Strategies,* 1(4): 87–99.

31. Molyboga, M. and C. L'Ahelec (2016). "A Simulation-Based Methodology for Evaluating Hedge Fund Investments." *Journal of Asset Management,* 17(6): 434–452.

32. Qian, E. (2006). "On the Financial Interpretation of Risk Contribution: Risk Budgets Do Add Up." *Journal of Investment Management,* 40(4): 1–11.

33. Maillard, S., Roncalli, T., and J. Teiletche (2010). "The Properties of Equally Weighted Risk Contribution Portfolios." *Journal of Portfolio Management,* 36(4): 60–70.

34. Clarke, R., De Silva, H., and S. Thorley (2013). "Risk Parity, Maximum Diversification, and Minimum Variance: An Analytic Perspective." *Journal of Portfolio Management,* 39(3): 39–53.

35. Qian, E. (2013). "Are Risk-Parity Managers at Risk Parity?" *Journal of Portfolio Management,* 40(1): 20–26.

36. Hallerbach, W. (2012). "A Proof of Optimality of Volatility Weighting over Time." *Journal of Investment Strategies,* 1(4): 87–99.

37. Molyboga, M. and C. L'Ahelec (2017). "Portfolio Management with Drawdown-Based Measures." *Journal of Alternative Investments,* 19(3): 75–89.

38. "Best Practices in Alternative Iinvestments: Due Diligence" (2010) by Greenwich Roundtable.

39. Rudin, A., Mor, V., and D. Farley (2020). "Adaptive Optimal Risk Budgeting." *Journal of Portfolio Management,* 46(4): 147–158.

40. Lee, W. and D.Y. Lam (2001). "Implementing Optimal Risk Budgeting." *Journal of Portfolio Management,* 28(1): 73–80.

41. Berkelaar, A.B., Kobor, A., and M. Tsumagari (2006). "The Sense and Nonsense of Risk Budgeting." *Financial Analysts Journal,* 62(5): 63–75.

42. Arnott, R, Campbell, R.H., Kalesnik, V., and J. Linnainmaa (2019). "Alice's Adventures in Factorland: Three Blunders That Plague Factor Investing." *Journal of Portfolio Management,* 45(4): 18–36.

43. Rudin, A. and D. Farley (2021). "Fuzzy Factors and Asset Allocation." *Journal of Portfolio Management,* 47(4): 110–122.

44. Barroso, P. and P. Santa-Clara (2015). "Momentum Has Its Moments." *Journal of Financial Economics,* 116(1): 111–120.

45. Daniel, T. and T.J. Moskowitz (2016). "Momentum Crashes." *Journal of Financial Economics,* 122(2): 221–247.

46. Kim, A.Y., Tse, Y., and J.K. Wald (2016). "Time Series Momentum and Volatility Scaling." *Journal of Financial Markets,* 30(3): 103–124.

47. Molyboga, M., Swedroe, L., and J. Qian (2020). "Short-Term Trend: A Jewel Hidden in Daily Returns." *Journal of Portfolio Management,* 47(1): 154–167.

48. Moreira, A. and T. Muir (2017). "Volatility-Managed Portfolios." *Journal of Finance,* 72(4): 1611–1643.

49. Cederburg, S., O'Doherty, M.S., Wang, F., and X.S. Yan (2020). "On the Performance of Volatility-Managed Portfolios." *Journal of Financial Economics,* 138(1): 95–117.

50. The 2020 study "On the Performance of Volatility-Managed Portfolios" criticized the 2017 paper "Volatility-Managed Portfolios" for its reliance on spanning regressions that are associated with structural instability and, thus, not implementable in real life.

51. Wang, F., Yan, X., and L. Zheng (2021). "Should Mutual Fund Investors Time Volatility?" *Financial Analysts Journal,* 77(1): 30–42.

52. Molyboga, M. (2019). "Portfolio Management of Commodity Trading Advisors with Volatility Targeting." *Journal of Investment Strategies,* 8(4): 1–20.

53. "RiskMetrics Technical Document, 4th edition" (1996).

54. Bollerslev, T. (1986). "Generalized Autoregressive Conditional Heteroskedasticity." *Journal of Econometrics,* 31(3): 307–327.

55. Pafka, S., Potters, M., and I. Kondor (2004). "Exponential Weighting and Random-Matrix-Theory-Based Filtering of Financial Covariance Matrices for Portfolio Optimization." *arXiv: Statistical Mechanics.*

Chapter 6. Advanced Portfolio Construction

1. Lopez de Prado, M. (2016). "Building Diversified Portfolios that Outperform Out of Sample." *Journal of Portfolio Management,* 42(4): 56–69.

2. Brinson, G.P., Hood, L.R., and G.L. Beebower (1995). "Determinants of Portfolio Performance." *Financial Analysts Journal,* 51(1): 133–138.

3. Anson, M. (2011). "The Evolution of Equity Mandates in Institutional Portfolios." *Journal of Portfolio Management,* 37(4): 127–137.

4. Ibid.

5. Lopez de Prado, M. (2016). "Building Diversified Portfolios that Outperform Out of Sample." *Journal of Portfolio Management,* 42(4): 56–69.

6. The 2016 paper "Building Diversified Portfolios that Outperform Out of Sample" includes a detailed description of the HRP approach and the Python implementation.

7. Molyboga, M. (2020). "A Modified Hierarchical Risk Parity Framework for Portfolio Management." *Journal of Financial Data Science,* 2(3): 128–139.

8. Molyboga, M. (2019). "Portfolio Management of Commodity Trading Advisors with Volatility Targeting." *Journal of Investment Strategies,* 8(4): 1–20.

9. Hallerbach, W. (2012). "A Proof of Optimality of Volatility Weighting over Time." *Journal of Investment Strategies,* 1(4): 87–99.

10. Lopez de Prado, M. (2016). "Building Diversified Portfolios that Outperform Out of Sample." *Journal of Portfolio Management,* 42(4): 56–69.

11. Ibid.

12. Ledoit, O. and L. Wolf (2017). "Nonlinear Shrinkage of the Covariance Matrix for Portfolio Selection: Markowitz Meets Goldilocks." *Review of Financial Studies,* 30(12): 4349–4388.

13. Ledoit, O. and L. Wolf (2003). "Improved Estimation of the Covariance Matrix of Stock Returns with an Application to Portfolio Selection." *Journal of Empirical Finance,* 10(5): 603–621.

14. Ledoit, O. and L. Wolf (2004). "Honey, I Shrunk the Sample Covariance Matrix." *Journal of Portfolio Management,* 30(4): 110–119.

15. Bun, J., Bouchaud, J.P., and M. Potters (2017). "Cleaning Large Correlation Matrices: Tools from Random Matrix Theory." *Physics Reports,* 666: 1–109.

16. Bun, J., Bouchaud, J.P., and M. Potters (2016). "Cleaning Correlation Matrices." *Risk.*

17. Ao, M., Yingying, L., and X. Zheng (2018). "Approaching Mean-Variance Efficiency for Large Portfolios." *Review of Financial Studies,* 32(7): 2890–2919.

18. Raponi, V., Uppal, R., and P. Zaffaroni (2021). "Robust Portfolio Choice." *SSRN working paper.*

19. Ao, M., Yingying, L., and X. Zheng (2018). "Approaching Mean-Variance Efficiency for Large Portfolios." *Review of Financial Studies,* 32(7): 2890–2919.

20. Kan, R. and G. Zhou (2007). "Optimal Portfolio Choice with Parameter Uncertainty." *Journal of Financial and Quantitative Analysis,* 42(3): 621–656.

21. Raponi, V., Uppal, R., and P. Zaffaroni (2021). "Robust Portfolio Choice." *SSRN working paper.*

22. Cooper, R.A. and M. Molyboga (2017). "Black-Litterman, Exotic Beta and Varying Efficient Portfolios: An Integrated Approach." *Journal of Investment Strategies,* 6(3): 1–18.

23. Roll, R. (1977). "A Critique of the Asset Pricing Theory's Tests Part I: On Past and Potential Testability of the Theory." *Journal of Financial Economics,* 4(2): 129–176.

24. DeMiguel, V., Garlappi, L., and R. Uppal (2009). "Optimal versus Naive Diversification: How Inefficient is the 1/N Portfolio Strategy?" *Review of Financial Studies,* 22(5): 1915–1953.

25. Asness, C., Frazzini, A., and L. Pedersen (2012). "Leverage Aversion and Risk Parity." *Financial Analysts Journal,* 68(1): 47–59.

26. Michaud, R.O. (1989). "The Markowitz Optimization Enigma: Is Optimized Optimal?" *Financial Analysts Journal,* 45(1): 31–42.

27. Cooper, R.A. and M. Molyboga (2017). "Black-Litterman, Exotic Beta and Varying Efficient Portfolios: An Integrated Approach." *Journal of Investment Strategies,* 6(3): 1–18.

28. Efron,B. (2005). "Bayesians, Frequentists, and Scientists." *Journal of the American Statistical Association,* 100: 1–5.

29. Michaud, R. and R. Michaud (2008). "Estimation Error and Portfolio Optimization: A Resampling Solution." *Journal of Investment Management,* 6(1): 8–28.

30. Markowitz, H. and N. Usmen (2003). "Resampled Frontiers versus Diffuse Bayes: An Experiment." *Journal of Investment Management,* 1(4): 9–25.

31. Harvey, C.R., Liechty, J.C., and M.W. Liechty (2008). "Bayes vs. Resampling: A Rematch." *Journal of Investment Management,* 6(1): 1–17.

32. "Milliman: Largest U.S. Public Pension Plans' funding ratio dips to 75% in August," Pensions and Investments, Rob Kozlowski, September 23, 2022.

33. "Commentary: Evaluation of Alternative Investments in Pension Fund Portfolios," Pensions and Investments, Marat Molyboga, October 31, 2019.

34. "Hedge Funds: Coping with Low Interest Rates," Investments & Pensions Europe, Michael Going and Marat Molyboga, March, 2021.

Chapter 7. Expert Hedge Fund Managers

1. Pedersen, L.H. (2015). "Efficiently Inefficient: How Smart Money Invests and Market Prices Are Determined." *Princeton University Press.*

Chapter 9. Inclusion and Diversity

1. Huang, R.R., Jie, E.Y., and Y. Ma (2022). "Life Cycle Performance of Hedge Fund Managers." *SSRN working paper.*

2. Getmansky Sherman, M., and H. Tookes (2022). "Female Representation in the Academic Finance Profession." *Journal of Finance,* 77(1): 317–365.

3. Billio, M., Lo, A., Pelizzon, L., Getmansky Sherman, M., and A. Zereei (2022). "Global Realignment in Financial Market Dynamics." *SSRN working paper.*

Appendix A. Manager Selection and Hedge Fund Factors

1. Molyboga, M., Baek, S., and J. Bilson (2017). "Assessing Hedge Fund Performance with Institutional Constraints: Evidence from CTA Funds." *Journal of Asset Management,* 18: 547–565.

2. Tibshirani, R. (1996). "Regression Shrinkage and Selection via the Lasso." *Journal of the Royal Statistical Society: Series B (Statistical Methodology),* 58(1): 267–288.

3. Zou, H. and T. Hastie (2005). "Regularization and Variable Selection via the Elastic Net." *Journal of the Royal Statistical Society: Series B (Statistical Methodology),* 67(2): 301–320.

4. Gu, S., Kelly, B., and D. Xiu (2020). "Empirical Asset Pricing via Machine Learning." *Review of Financial Studies,* 33(5): 2223–2273.

Appendix B. Performance Persistence

1. Kosowski, R., Timmermann, A., Wermers, R., and H. White (2006). "Can Mutual Fund 'Stars' Really Pick Stocks? New Evidence from a Bootstrap Analysis." *Journal of Finance,* 61(6): 2551–2595.

2. Kosowski, R., Naik, N.Y., and M. Teo (2007). "Do Hedge Funds Deliver Alpha? A Bayesian and Bootstrap Analysis." *Journal of Financial Economics,* 84(1): 229–264.

3. Fama, E.F. and K.R. French (2010). "Luck versus Skill in the Cross-Section of Mutual Fund Returns." *Journal of Finance,* 65(5): 1915–1947.

4. Ibid.

5. Kosowski, R., Timmermann, A., Wermers, R., and H. White (2006). "Can Mutual Fund 'Stars' Really Pick Stocks? New Evidence from a Bootstrap Analysis." *Journal of Finance,* 61(6): 2551–2595.

6. Ibid.

7. Fama, E.F. and K.R. French (2010). "Luck versus Skill in the Cross-Section of Mutual Fund Returns." *Journal of Finance,* 65(5): 1915–1947.

8. Kosowski, R., Naik, N.Y., and M. Teo (2007). "Do Hedge Funds Deliver Alpha? A Bayesian and Bootstrap Analysis." *Journal of Financial Economics,* 84(1): 229–264.

9. Barras, L., Scaillet, O., and R. Wermers (2010). "False Discoveries in Mutual Fund Performance: Measuring Luck in Estimated Alphas." *Journal of Finance,* 65(1): 179–216.

10. Storey, J.D. (2002). "A Direct Approach to False Discovery Rates." *Journal of the Royal Statistical Society,* 64: 479–498.

11. Barras, L., Scaillet, O., and R. Wermers (2010). "False Discoveries in Mutual Fund Performance: Measuring Luck in Estimated Alphas." *Journal of Finance,* 65(1): 179–216.

12. Andrikogiannopoulou, A. and F. Papakonstantinou (2019). "Reassessing False Discoveries in Mutual Fund Performance: Skill, Luck, or Lack of Power?" *Journal of Finance,* 74(5): 2667–2688.

13. Barras, L., Scaillet, O., and R. Wermers (2010). "False Discoveries in Mutual Fund Performance: Measuring Luck in Estimated Alphas." *Journal of Finance,* 65(1): 179–216.

14. Andrikogiannopoulou, A. and F. Papakonstantinou (2019). "Reassessing False Discoveries in Mutual Fund Performance: Skill, Luck, or Lack of Power?" *Journal of Finance,* 74(5): 2667–2688.

15. Ibid.

16. Harvey, C.R. and Y. Liu (2018). "Detecting Repeatable Performance." *Review of Financial Studies,* 31(7): 2499–2552.

17. Pastor, L. and R.F. Stambaugh (2002). "Mutual Fund Performance and Seemingly Unrelated Assets." *Journal of Financial Economics,* 63(3): 315–349.

18. Forsberg, D., Gallagher, D.R., and G.J. Warren (2021). "Identifying Hedge Fund Skill by Using Peer Cohorts." *Financial Analysts Journal,* 77(2): 97–123.

19. Agarwal, V., Daniel, N.D., and N.Y. Naik (2009). "Role of Managerial Incentives and Discretion in Hedge Fund Performance." *Journal of Finance,* 64(5): 2221–2256.

Appendix C. From Mean Variance to Risk Parity

1. Stein, C. (1956). "Inadmissibility of the Usual Estimator for the Mean of a Multivariate Normal Distribution." *Proceedings of the Third Berkeley Symposium on Mathematical Statistics and Probability,* 1: 197–206.

2. Jorion, P. (1986). "Bayes-Stein Estimation for Portfolio Analysis." *Journal of Financial and Quantitative Analysis,* 21(3): 279–292.

3. Merton, R.C. (1980). "On Estimating the Expected Return on the Markets: An Exploratory Investigation." *Journal of Financial Economics,* 8(4): 323–361.

4. Ledoit, O. and L. Wolf (2003). "Improved Estimation of the Covariance Matrix of Stock Returns with an Application to Portfolio Selection." *Journal of Empirical Finance,* 10(5): 603–621.

5. Ledoit, O. and L. Wolf (2004). "Honey, I Shrunk the Sample Covariance Matrix." *Journal of Portfolio Management,* 30(4): 110–119.

6. Sharpe, W.F. (1963). "A simplified model for portfolio analysis." *Management Science,* 9(1): 277–293.

7. Yang, Z., Han, K., Molyboga, M., and G. Molyboga (2016). "A New Diagnostic Approach to Evaluating the Stability of Optimal Portfolios." *Journal of Investing,* 25(1): 37–45.

8. Ledoit, O. and L. Wolf (2017). "Nonlinear Shrinkage of the Covariance Matrix for Portfolio Selection: Markowitz Meets Goldilocks." *Review of Financial Studies,* 30(12): 4349–4388.

9. Ledoit, O. and L. Wolf (2003). "Improved Estimation of the Covariance Matrix of Stock Returns with an Application to Portfolio Selection." *Journal of Empirical Finance,* 10(5): 603–621.

10. Ledoit, O. and L. Wolf (2004). "Honey, I Shrunk the Sample Covariance Matrix." *Journal of Portfolio Management,* 30(4): 110–119.

11. Ledoit, O. and L. Wolf (2022). "The Power of (Non-)Linear Shrinking: A Review and Guide to Covariance Matrix Estimation." *Journal of Financial Econometrics,* 20(1): 187–218.

12. Ledoit, O. and L. Wolf (2021). "Shrinkage Estimation of Large Covariance Matrices: Keep It Simple, Statistician?" *Journal of Multivariate Analysis,* 186.

13. Black, F. and R. Litterman (1992). "Global Portfolio Optimization." *Financial Analysts Journal,* 48(5): 28–43.

14. Merton, R.C. (1980). "On Estimating the Expected Return on the Markets: An Exploratory Investigation." *Journal of Financial Economics,* 8(4): 323–361.

15. DeMiguel, V., Garlappi, L., and R. Uppal (2009). "Optimal versus Naive Diversification: How Inefficient is the 1/N Portfolio Strategy?" *Review of Financial Studies,* 22(5): 1915–1953.

16. Kan, R. and G. Zhou (2007). "Optimal Portfolio Choice with Parameter Uncertainty." *Journal of Financial and Quantitative Analysis,* 42(3): 621–656.

17. Hallerbach, W. (2012). "A Proof of Optimality of Volatility Weighting over Time." *Journal of Investment Strategies,* 1(4): 87–99.

18. Qian, E. (2006). "On the Financial Interpretation of Risk Contribution: Risk Budgets Do Add Up." *Journal of Investment Management,* 40(4): 1–11.

19. Artzner, P., Delbaen, F., Eber, J.M., and D. Heath (1999). "Coherent Measures of Risk." *Mathematical Finance,* 9(3): 203–228.

20. "Best Practices in Alternative Investments: Due Diligence" (2010) by Greenwich Roundtable.

21. Michaud, R.O. (1989). "The Markowitz Optimization Enigma: Is Optimized Optimal?" *Financial Analysts Journal,* 45(1): 31–42.

22. Jorion, P. (1986). "Bayes-Stein Estimation for Portfolio Analysis." *Journal of Financial and Quantitative Analysis,* 21(3): 279–292.

23. DeMiguel, V., Garlappi, L., and R. Uppal (2009). "Optimal versus Naive Diversification: How Inefficient is the 1/N Portfolio Strategy?" *Review of Financial Studies,* 22(5): 1915–1953.

24. Douady, R., Shiryaev, A., and M. Yor (2000). "On Probability Characteristics of Downfalls in a Standard Brownian Motion." *Theory of Probability and Its Applications,* 44(1): 29–38.

25. Chekhlov, A., Uryasev, S., and M. Zabarankin (2005). "Drawdown Measure in Portfolio Optimization." *International Journal of Theoretical and Applied Finance,* 8(1): 13–58.

26. Goldberg, L. and O. Mahmoud (2014). "On a Convex Measure of Drawdown Risk." *working paper.*

27. Molyboga, M. and C. L'Ahelec (2017). "Portfolio Management with Drawdown-Based Measures." *Journal of Alternative Investments,* 19(3): 75–89.

28. Rudin, A., Mor, V., and D. Farley (2020). "Adaptive Optimal Risk Budgeting." *Journal of Portfolio Management,* 46(4): 147–158.

29. Lee, W. and D.Y. Lam (2001). "Implementing Optimal Risk Budgeting." *Journal of Portfolio Management,* 28(1): 73–80.

30. Berkelaar, A.B., Kobor, A., and M. Tsumagari (2006). "The Sense and Nonsense of Risk Budgeting." *Financial Analysts Journal,* 62(5): 63–75.

31. Molyboga, M. (2019). "Portfolio Management of Commodity Trading Advisors with Volatility Targeting." *Journal of Investment Strategies,* 8(4): 1–20.

32. Merton, R.C. (1980). "On Estimating the Expected Return on the Markets: An Exploratory Investigation." *Journal of Financial Economics,* 8(4): 323–361.

33. Bollerslev, T. (1986). "Generalized Autoregressive Conditional Heteroskedasticity." *Journal of Econometrics,* 31(3): 307–327.

34. Harvey, C.R. (2001). "The Specification of Conditional Expectations." *Journal of Empirical Finance,* 8(5): 573–637.

35. Ang, A. and J. Liu (2007). "Risk, Return and Dividends." *Journal of Financial Economics,* 85(1): 1–38.

Appendix D. Advanced Portfolio Construction

1. Bun, J., Bouchaud, J.P., and M. Potters (2016). "Cleaning Correlation Matrices." *Risk.*

2. Merton, R.C. (1980). "On Estimating the Expected Return on the Markets: An Exploratory Investigation." *Journal of Financial Economics,* 8(4): 323–361.

3. Ledoit, O. and L. Wolf (2003). "Improved Estimation of the Covariance Matrix of Stock Returns with an Application to Portfolio Selection." *Journal of Empirical Finance,* 10(5): 603–621.

4. Ao, M., Yingying, L., and X. Zheng (2018). "Approaching Mean-Variance Efficiency for Large Portfolios." *Review of Financial Studies,* 32(7): 2890–2919.

5. Kan, R. and G. Zhou (2007). "Optimal Portfolio Choice with Parameter Uncertainty." *Journal of Financial and Quantitative Analysis,* 42(3): 621–656.

6. Efron, B., Hastie, T., Johnstone, J., and R. Tibshirani (2004). "Least Angle Regression." *Annals of Statistics,* 32(2): 407–499.

7. Raponi, V., Uppal, R., and P. Zaffaroni (2021). "Robust Portfolio Choice." *SSRN working paper.*

8. Hansen, L.P. and T.J. Sargent (2007). "Robustness." *Princeton University Press.*

9. Cooper, R.A. and M. Molyboga (2017). "Black-Litterman, Exotic Beta and Varying Efficient Portfolios: An Integrated Approach." *Journal of Investment Strategies,* 6(3): 1–18.

Index

Note: Locators in *italics* represent figures and **bold** indicate tables in the text.

J

K

L

Printed in the United States
by Baker & Taylor Publisher Services